知识生产的原创基地
BASE FOR ORIGINAL CREATIVE CONTENT
颉腾科技
JIE TENG TECHNOLOGY

CiaoCiao 船长
征服Java

喂养大脑更好的Java技能练习

卷 II

Captain CiaoCiao
erobert Java

[德] 克里斯蒂安·尤伦布姆 / 著
徐承志 / 译

北京理工大学出版社
BEIJING INSTITUTE OF TECHNOLOGY PRESS

版权专有　侵权必究

图书在版编目（CIP）数据

CiaoCiao 船长征服 Java：喂养大脑更好的 Java 技能练习卷. Ⅱ /（德）克里斯蒂安·尤伦布姆著；徐承志译. -- 北京：北京理工大学出版社，2025.6.

ISBN 978-7-5763-5425-6

Ⅰ. TP312.8

中国国家版本馆 CIP 数据核字第 2025WK5039 号

北京市版权局著作权合同登记号　图字：01-2023-3102 号
Copyright © 2021 Rheinwerk Verlag, Bonn 2021. All rights reserved.
First published in the German language under the title "Captain CiaoCiao erobert Java"（ISBN 978-3-8362-8427-1）by Rheinwerk Verlag GmbH, Bonn, Germany.
Simplified Chinese edition copyright © 2025 by Beijing Jie Teng Culture Media Co., Ltd. through Media Solutions, Tokyo Japan（info@mediasolutions.jp）
All rights reserved. Unauthorized duplication or distribution of this work constitutes copyright infringement.

责任编辑：钟　博	**文案编辑**：钟　博
责任校对：刘亚男	**责任印制**：施胜娟

出版发行	/ 北京理工大学出版社有限责任公司
社　　址	/ 北京市丰台区四合庄路 6 号
邮　　编	/ 100070
电　　话	/（010）68944451（大众售后服务热线）
	（010）68912824（大众售后服务热线）
网　　址	/ http://www.bitpress.com.cn

版 印 次	/ 2025 年 7 月第 1 版第 1 次印刷
印　　刷	/ 三河市中晟雅豪印务有限公司
开　　本	/ 787 mm×1092 mm　1/16
印　　张	/ 23.75
字　　数	/ 410 千字
定　　价	/ 169.00 元

图书出现印装质量问题，请拨打售后服务热线，负责调换

亲爱的读者朋友：

很荣幸在此介绍海盗 Bonny Brain、船长 CiaoCiao 和各位船员。海盗们为自己设定了一个目标：征服 Java。他们不仅想在此登陆，更是想要最终占领 Java 中每一处有趣的角落。

如果你也有同样的目标，那你就找对地方啦。或许你听过本书作者克里斯蒂安·尤伦布姆（Christian Ullenboom）的另一本书：《Java 岛：程序员经典标准教程》(*Java ist auch eine Insel. Einführung, Ausbildung, Praxis*)，该书包罗万象，是一本非常详尽的综合手册。而在本书中你也将受益于作者丰富的知识储备和讲师经验，书中包含了他实操多年的练习材料。

你可以将本书与任一教材或其他资料配合使用，因为在每章的开头都有该章练习内容所属类别和方法的摘要。相对应地，你也能够很快地为每个主题找到合适的练习。

本书最为便捷的使用方式是配合作者所编写的综合手册进行练习：你可以通过练习里的章节指示快速地找到其在《Java 岛：程序员经典标准教程》中的相应内容。本练习手册对应《Java 岛：程序员经典标准教程》的原书第 15 版，但你也可以通过标题指引配合其他版次的综合手册进行学习。有时，本手册还会提到《Java SE 9 标准库：Java 开发人员手册》(*Java SE 9 Standard-Bibliothek. Das Handbuch für Java-Entwickler*) 中的章节内容以便深入学习。

本书适用于自 Java 8 起的所有 Java 版本。如果任务涉及 Java 的更高版本，则会明确说明它是属于哪个功能和自第几版本起的 Java 可以使用。

小提示：如果你想尽办法，但 Java 仍无法按预期运行，或者你有任何建议，请随时与我们联系。期待你的建设性批评！

祝你和船长及他的船员们玩得开心！

Rheinwerk Computing 的编辑阿尔穆特·波尔（Almut Poll）

可以通过以下方式联系我
电子邮件：almut.poll@rheinwerk-verlag.de
网站：www.rheinwerk-verlag.de
地址：Rheinwerk Verlag·Rheinwerkallee 4·53227 Bonn

目录
CONTENTS

序言

第 1 章　空间与时间　　　　　　　　　　　　　　　　　001

1.1　语言和国家　　　　　　　　　　　　　　　　　　　002
1.1.1　针对随机数应用特定国家 / 语言的格式化★　　　002
1.2　日期类和时间类　　　　　　　　　　　　　　　　　003
1.2.1　以不同语言格式化日期输出★　　　　　　　　003
1.2.2　今年弗朗西斯·博福特爵士的生日是星期几？★　004
1.2.3　确定卡拉 OK 派对的平均时长★　　　　　　　004
1.2.4　解析不同的日期格式★★★　　　　　　　　　005
1.3　可供参考的解决方案　　　　　　　　　　　　　　　006

第 2 章　基于线程的并发编程　　　　　　　　　　　　　014

2.1　创建线程　　　　　　　　　　　　　　　　　　　　015
2.1.1　创建用于招手和挥动旗帜的线程★　　　　　　016
2.1.2　不再招手或挥动旗帜：结束线程★　　　　　　017
2.1.3　参数化 Runnable ★★　　　　　　　　　　　017
2.2　执行和休眠　　　　　　　　　　　　　　　　　　　018
2.2.1　使用睡眠线程延迟处理★★　　　　　　　　　018
2.2.2　通过线程观察文件变化　　　　　　　　　　　019
2.2.3　捕捉异常★　　　　　　　　　　　　　　　　019

2.3	线程池和结果	020
2.3.1	使用线程池 ★★	021
2.3.2	确定网站的最后修改 ★★	021
2.4	保护临界区	023
2.4.1	在诗集里写下回忆 ★	024
2.5	线程协作和同步助手	026
2.5.1	与船长一起参加宴会——Semaphore ★★	027
2.5.2	咒骂与侮辱——Condition ★★	027
2.5.3	从颜料盒中取出笔——Condition ★★	027
2.5.4	玩"剪刀石头布"游戏——CyclicBarrier ★★	028
2.5.5	找到跑得最快的人——CountDownLatch ★★	030
2.6	可供参考的解决方案	030

第 3 章　数据结构和算法　　055

3.1	集合 API 的接口	056
3.1.1	测试：搜索 StringBuilder ★★	058
3.2	列表	059
3.2.1	唱歌做饭：运行列表，检查特性 ★	059
3.2.2	从列表中过滤评论 ★	060
3.2.3	缩短列表，因为衰退不存在 ★	061
3.2.4	和朋友一起吃饭：比较元素，找到共同点 ★	062
3.2.5	检查列表的相同元素顺序 ★	062
3.2.6	现在播报天气：寻找重复元素 ★	063
3.2.7	创建收据输出 ★	064
3.2.8	测试：装饰数组 ★	065
3.2.9	测试：查找和未找到 ★	065
3.2.10	加上奶酪，一切都会更美味：在列表中添加元素 ★	065
3.2.11	测试：一无所有让人很恼火 ★	065
3.2.12	使用迭代器搜索元素，找到 Covid Cough ★★	066
3.2.13	移动元素，玩"抢椅子"游戏 ★	067
3.2.14	编辑行星的问答游戏 ★	068
3.3	集合	069

3.3.1	形成子集，寻找共同点 ★	070
3.3.2	测试：好剑 ★	070
3.3.3	删除数组中的重复元素 ★	071
3.3.4	查明单词中包含的所有单词 ★★	072
3.3.5	正确分类几乎相同的东西 ★★	073
3.3.6	用 UniqueIterator 排除重复元素 ★★	073
3.4	关联映射	074
3.4.1	将二维数组转换为映射 ★	074
3.4.2	将文本转换为摩尔斯密码并反转 ★	075
3.4.3	用关联映射标记词频 ★★	075
3.4.4	读取颜色并播放 ★★	076
3.4.5	读取名称，管理长度 ★★	077
3.4.6	找到缺失的字符 ★★	078
3.4.7	计算寻找三头猴的路径数 ★★	078
3.4.8	在排序的关联映射中管理节日 ★	080
3.4.9	测试：HashMap 中的值 ★★	081
3.4.10	确定共同点：派对场所布置和带来的礼物 ★	081
3.5	Properties	082
3.5.1	开发便捷的属性装饰器 ★★	082
3.6	堆栈 (Stack) 和队列 (Queue)	084
3.6.1	编辑 RPN（逆波兰表示法）计算器 ★	084
3.7	BitSet	085
3.7.1	查找重复条目并解决动物混乱 ★	085
3.8	线程安全的数据结构	086
3.8.1	装船 ★★	086
3.8.2	优先编辑重要消息 ★	087
3.8.3	用完就换新的 ★★★	089
3.9	建议解决方案	090

第 4 章　Java Stream API　　　　　　　　　　　　　　　150

4.1	常规流及其终端和中间操作	151
4.1.1	超级英雄史诗：认识 Stream API ★	151

4.1.2	测试：双倍输出 ★	152
4.1.3	从列表中找出心爱的船长 ★	152
4.1.4	框住图片（Java 11）★	152
4.1.5	看和说 ★★	154
4.1.6	删除包含稀土金属的重复岛屿（Java 9）★★★	155
4.1.7	船帆在哪里？★★	156
4.1.8	购买最受欢迎的装甲车 ★★★	157
4.2	**原始流**	**158**
4.2.1	识别数组中的非数字 ★	158
4.2.2	生成数十年 ★	158
4.2.3	通过 Stream 创建具有恒定内容的数组 ★	159
4.2.4	绘制金字塔（Java 11）★	159
4.2.5	查找字符串的字母频率 ★	159
4.2.6	从 1 到 0，从 10 到 9 ★★	160
4.2.7	合并两个 int 数组 ★★	160
4.2.8	确定获胜组合 ★★	161
4.3	**统计数据**	**161**
4.3.1	最快和最慢的桨手 ★	162
4.3.2	计算中位数 ★★	163
4.3.3	统计温度并绘制图表 ★★★	163
4.4	**可供参考的解决方案**	**165**

第 5 章　文件和对文件内容的随机访问　　195

5.1	**路径和文件**	**196**
5.1.1	显示每日格言 ★	196
5.1.2	合并隐藏 ★	196
5.1.3	创建文本副本 ★★	197
5.1.4	生成目录列表 ★	197
5.1.5	搜索大型 GIF 文件 ★	198
5.1.6	递归降级目录并找到空的文本文件 ★	199
5.1.7	开发自己的文件过滤工具库 ★★★	199
5.2	**对文件内容的随机访问**	**200**

5.2.1	输出文本文件的最后一行★	200
5.3	可供参考的解决方案	200

第 6 章　输入 / 输出　　214

6.1	直接数据流	216
6.1.1	确定不同位置的数量（读取文件）	216
6.1.2	将 Python 程序转换为 Java 程序（文件写入）★	216
6.1.3	生成目标代码（写入文件）★	217
6.1.4	将文件内容转换为小写字母（文件的读取和写入）	219
6.1.5	以 ASCII 灰度显示 PPM 图像★★★	219
6.1.6	文件分块处理（文件的读和写）	221
6.2	嵌套流	222
6.2.1	测试 DataInputStream 和 DataOutputStream	222
6.2.2	使用 GZIPOutputStream 压缩数字序列 ★	222
6.3	序列化	222
6.3.1	对聊天的数据进行（反）序列化并转换为文本★★	223
6.3.2	测试：序列化的前提条件★	223
6.3.3	保存最后的条目 ★★	223
6.4	建议解决方案	224

第 7 章　网络编程　　245

7.1	URL 和 URL 连接	245
7.1.1	通过 URL 下载远程图像	246
7.1.2	通过 URL 读取远程文本文件★	246
7.2	HTTP 客户端（Java 11）	247
7.2.1	来自 Hacker News 的头条	248
7.3	套接字和服务器套接字	248
7.3.1	运行一个演讲服务器和它的客户端	249
7.3.2	实现一个端口的扫描器★★	249
7.4	建议解决方案	251

第 8 章　用 Java 处理 XML、JSON 等数据格式的文件　　262

8.1　用 Java 处理 XML 文件　　263
8.1.1　编写带有配方的 XML 文件　　263
8.1.2　检查所有图片是否有一个 alt 属性　　264
8.1.3　编写带有 JAXB 的 Java 对象　　265
8.1.4　阅读笑话并开心地笑★★　　266
8.2　JSON　　268
8.2.1　黑客新闻：评估 JSON 文件　　269
8.2.2　以 JSON 格式读取写入编辑器的配置★★　　270
8.3　HTML　　271
8.3.1　使用 jsoup 加载维基百科图像★★　　271
8.4　Office 文件　　271
8.4.1　生成带截图的 Word 文件　　271
8.5　归档　　272
8.5.1　播放 ZIP 档案中的昆虫声音★★　　272
8.6　建议解决方案　　273

第 9 章　用 JDBC 访问数据库　　290

9.1　数据库管理系统　　291
9.1.1　准备 H2 ★　　291
9.2　数据库查询　　291
9.2.1　查询所有注册的 JDBC 驱动程序 ★　　291
9.2.2　建立数据库并执行 SQL 脚本★　　292
9.2.3　向数据库写入数据★　　294
9.2.4　在批处理模式下向数据库添加数据★　　294
9.2.5　用准备好的指令添加数据★　　295
9.2.6　数据查询　　295
9.2.7　以交互方式滚动 ResultSet ★　　296
9.2.8　海盗存储库　　296
9.2.9　查询列元数据　　298
9.3　建议解决方案　　298

第 10 章　操作系统的接口　　313

10.1　控制台　　314
10.1.1　彩色的控制台输出 ★　　314
10.2　特性　　315
10.2.1　Windows 或者 UNIX 或者 macOS?　　315
10.2.2　统一命令行特性和文件中的特性　　316
10.3　运行外部程序　　316
10.3.1　通过 Windows Management Instrumentation 读取电池状态　　316
10.4　建议解决方案　　317

第 11 章　反射、注解和 JavaBeans　　326

11.1　反射 API　　326
11.1.1　创建具有继承关系的 UML 图示 ★　　327
11.1.2　创建带有属性的 UML 图示 ★　　328
11.1.3　从清单条目中生成 CSV 文件　　329
11.2　注释　　330
11.2.1　从注释的对象变量中创建 CSV 文件 ★★　　330
11.3　建议解决方案　　331

编后语　　341

附录 A　Java 领域中常见的类型和方法　　343

A.1　经常出现的类型的包　　343
A.2　100 个最常使用的类　　345
A.3　100 种常用的方法　　349
A.4　100 个最常用的方法，包括参数列表　　353

序言
PREFACE

刚开始接触编程的人会不断地问自己同样的问题：作为开发人员，我该如何取得进步，如何更好地编程？答案其实很简单：阅读、观看网络视频、学习、复习、练习、讨论。学习编程与学习其他技能的许多方面都有相通之处。只靠读书无法学会演奏一种乐器，观看《速度与激情》系列电影也不能教会我们开车。只有不断地练习和重复才能在大脑中建立模式和结构。学习自然语言和学习编程语言之间有很大的相似之处。坚持使用语言，以及使用语言表达自己和进行交流的迫切渴望与急切要求，例如使用这门语言去购买汉堡或啤酒，可以促使技能得到稳步提高。

市面上并不缺乏学习编程语言的书籍和网络视频。然而，阅读、学习、练习和重复只是第一步。我们必须把知识创造性地结合起来，才能成功地开发软件解决方案。而这一点只有通过练习才能做到，就像音乐家经常做手指练习和演奏曲目一样。练习效果越好，效率越高，就能越快成为大师。本书旨在帮助你迈出下一步并获得更多实践经验。

Java 16 声明包含大约 4 400 个类、1 300 个接口，以及更多的列表、注释和异常。这些类型中只有一小部分与实践相关，本书选择了其中最重要的类型和方法并提供相应的练习。我竭尽所能地将书中的练习编写得生动有趣，并提供符合 Java 编码规范的示例性解决方案，此外，也会展示多种多样的替代解决方案和方法。希望我所提出的解决方案能清楚地展示非功能性需求，因为程序的质量不仅体现在"做它应该做的"，诸如正确缩进、遵守命名规范、正确使用修饰符、投入最佳实践、设计模式等也很重要，这些都是本书中提供的解决方案应展示的内容。简而言之：干净的代码。

知识储备和目标群体

本书专为初次接触 Java 或已经进阶并希望继续学习 Java SE 标准库的 Java 开

发人员设计，主要面向以下目标群体：

- 计算机专业大学生；
- IT 专家；
- Java 程序员；
- 软件开发人员；
- 求职者。

本书的重点在于练习任务和完整的解决方案提议，其中包含有关 Java 特点、优秀的面向对象编程、最佳实践和设计模式的详细背景信息。本书中的练习任务最好配合教科书使用，因为本书相当于一本任务集，并不是一本典型的教科书。你可以配套使用我所编写的手册《Java 岛：程序员经典标准教程》——在本书中我标记了各个任务在手册中的相应章节，你也可以使用其他教科书，先学习所选教科书中的一个主题，然后解决本书中相应的任务，这也不失为一个好的学习方法。

本书中的前期任务针对刚开始使用 Java 的编程初学者，而学习 Java 的时间越长，相应的 Java 任务要求就越高。故在书中对于初学者和高级开发人员都有相应水平的练习任务。

Java 标准版可被许多框架和库扩展。本任务集不涉及特定库或 Java Enterprise 框架，如 Jakarta EE 或 Spring（Boot）。市场上有针对这种环境单独的练习册。本书也不需要 Profiling 之类的工具。

使用说明

本书分为不同的模块，其中包含针对 Java 语言、Java 标准库的精选内容，如数据结构或文件处理等相关练习。每个模块都包含编程任务，此外还增加了附带小惊喜的"测验"。每一部分的内容都从学习动机、主题分类导入，紧接着是相关的练习任务。如果任务特别棘手，还会有额外的提示。其他任务可通过自行选择决定是否继续拓展练习。

绝大多数的任务之间并没有直接的关联，读者可以从任意一个任务入手进行练习。只有在命令式编程章节，部分练习包含相对较大的项目，以及在面向对象编程章节中，部分任务是相互关联的。你可以在任务描述中找到相关说明。相对复杂的程序不仅能够帮助你结合任务学习不同的语言属性，赋予整个项目更多实际意义，还能激励你继续学习。

每项任务都会附有本人对任务复杂度的评估，以 1 星、2 星或 3 星来标记。当然，你在做任务过程中，也许对任务的难度会有不同的感受。

1 星★：任务简单，适合初学者。它们很容易解决，不是很费力，通常只需要迁移知识，如转写教科书中的内容。

2 星★★：需要付出更多时间，需要将不同的技术结合起来，需要拥有更强的创造力。

3 星★★★：任务更复杂，需要更多的知识积累，有时还需要检索相关资料。这些任务通常无法通过单一方法解决，而需要多个协同工作的类。

解决方案建议

每项任务至少有一个建议的解决方案。我不想称之为"标准答案"，以免暗示给定的解决方案是最好的，而所有其他解决方案都是无用的。读者应该将自己的解决方案与建议的解决方案进行比较，如果自己的解决方案比给出的解决方案更清晰，这是值得高兴的事情。书中提出的解决方案附带注解，因此理解各步骤应该不成问题。

为了避免大家在阅读任务后就直接翻看答案，所有建议的解决方案都收录在每一章节的结尾。这样自己解决任务的乐趣就不会被剥夺。

你可以在 https://www.rheinwerk-verlag.de/5329/ 和 https://github.com/ullenboom/captain-ciaociao 这两个网站上找到所有建议的解决方案。在一些解决方案中会出现 //tag::solution[ ] 样式的注释，它表明有关这些答案的相关段落是在本书中出现过的。

结合教科书和本书学习 Java

本书是作为 Java 教科书（或视频课程）的扩展实践部分而编写的。虽然本书并不是其他教科书的配套练习书或实践书，但本书内容与作者所著的下列两本教科书的顺序相同：

▶ 《Java 岛：程序员经典标准教程》（Java 岛 1）
▶ 《Java SE 9 标准库：Java 开发人员手册》（Java 岛 2）

本书的章节开头附有《Java 岛：程序员经典标准教程》的相应章节指引，书中几乎所有的练习都可以通过学习《Java 岛：程序员经典标准教程》来解决，但并不是《Java 岛：程序员经典标准教程》的每一章节都能在本书中找到练习任务，如本书中并没有关于图形界面的内容。

本书优势

每个软件开发人员的目标都是将任务转换为程序，而这需要大量的练习和范例。本书很好地结合了两者。网上固然也有非常多的任务，但这些任务往往是混杂的，不成体系，其解决方案可能也已经过时。本书系统地编制任务，并提供了清晰的解决方案建议。研究这些建议解决方案并阅读代码通常有助于记住一些特定的模型。无法想象不阅读就能研读圣经，然而，软件开发人员阅读的第三方代码相对较少，他们往往只生产自己的代码，但讽刺的是，一段时间过后他们很可能就无法理解自己写过的代码了。阅读也可以促进写作。阅读时，我们的大脑会存储一些模型和解决方案，而这些模型和方案会被我们下意识地运用到我们自己的代码写作中。大脑是一个奇妙的器官，它可以根据模型自动联想，从中学习并进行加工。输入大脑的模型质量越好，神经结构就会越好。糟糕的解决方案是不良范例，我们应当只用好的代码"喂食"我们的大脑。以异常处理或错误处理为例，本书的所有建议解决方案都讨论了正确的输入值、可能发生的错误状态以及如何处理这些错误状态。我们并非生活在一个完美世界中，软件中总是会出现各种各样的问题，对此我们必须做好准备，不能忽视不完美的世界。

许多开发人员受限于自己编写软件的方式。因此，阅读新的解决方案以扩充词汇量是非常必要的。Java 开发人员的词汇是库。在企业工作的许多 Java 开发人员极少以面向对象的方式进行编程，而是编写庞大的类。本书旨在提高面向对象编程建模领域的技能。本书引入了新方法，创建了新数据类型，将复杂性降至最低。此外函数式编程在当前的 Java 编程中也扮演着重要的角色。所有解决方案始终基于 Java 8（及更高版本）的现代语言。

一些建议解决方案可能过于复杂，但任务和建议解决方案可以帮助开发人员学会专注并反思思考步骤。专注力和快速掌握代码的能力在实践中非常重要，因为开发人员经常需要加入一个新的团队并能够理解和扩展其他源代码，甚至修复程序漏洞。即使是想要扩展现有开源解决方案的人也可以从这种专注练习中受益。

除了重点介绍 Java 编程语言、语法、库和面向对象外，本书还提供了大量关于算法、历史发展、与其他编程语言和数据格式的比较等内容。

如果还需要一个必须拥有本书的理由，那我告诉你，它也可以助你入眠。

所需软件

原则上，每项任务都可以用纸笔解决。如果回到 50 年前，这甚至是唯一的解决方法。但现如今专业的软件开发需要工具并需要能够正确使用工具。编程语言的句法知识、面向对象建模的可能性和标准库只是一方面，另一方面也需要 JVM（Java 虚拟机）、Maven 等工具或 Git 等版本控制系统，以及开发环境。有些开发者可以在开发环境中施展魔术，神奇地生成源代码，自动修正错误。

JVM

运行 Java 程序需要 JVM。之前这并不难，因为运行时环境最初由美国太阳微系统公司（Sun）提供。然而太阳微系统公司之后被甲骨文公司（Oracle）所收购，我们虽仍然可从甲骨文公司获得运行时环境，但许可条款已被更改。我们可以使用甲骨文公司的 Java 软件开发工具包（Oracle JDK）进行软件测试和开发，但不能进行生产操作。如需要进行此类操作，甲骨文公司要求收取专利许可费。这导致很多机构使用开源版本的 Java 软件开发工具包（OpenJDK）编译自己的运行时环境。读者可以使用 Eclipse Adoptium（仍以 AdoptOpenJDK 的名称为人所知，https://adoptopenjdk.net）、Amazon Corretto（https://aws.amazon.com/de/corretto）、Red Hat OpenJDK（https://developers.redhat.com/products/openjdk/overview）和其他产品，如来自 Azul Systems 或 Bellsoft 的产品编译运行时环境。读者可以选择任意一个发行版，没有强制规定。

本书中的大部分任务都可以使用 Java 8 版本解决，无须使用更高版本的 Java。当然，新的 Java 版本提供了语言和库的扩展，但 Java 8 在企业中仍被广泛使用。具有更新语言元素和应用程序编程接口（API）的任务带有标识符，读者可以自行筛选任务。

开发环境

Java 源代码只是文本，因此原则上一个简单的文本编辑器就足够了。但是，我们不能期望像记事本这样的编辑器具有很高的生产力。现代开发环境可以支持我们完成许多任务：关键字的颜色突出显示、自动完成代码、智能纠错、插入代码块、可视化调试器中的状态等。因此，建议读者使用完整的开发环境。目前流行的四种集成开发环境（IDE）是 IntelliJ，Eclipse，Visual Studio Code 和 (Apache) NetBeans。与 Java 运行时环境一样，读者可以自行选择开发环境。Eclipse，NetBeans 和 Visual Studio Code 是免费开源的，IntelliJ 社区版也是免费开源的，但功能更强大的 IntelliJ 终极版是收费的。

自本书中间的某些任务开始，需要通过 Maven 来实现项目依赖关系。

本书代码格式说明

本书中代码使用等宽字体，文件名使用斜体。为了区分方法和属性，方法总是包含一对括号，如"变量 max 包含最大值"或"max( ) 返回最大值"。由于方法可以重载，所以在命名参数列表时，要么如 equals (Object) 这样，要么用省略号缩写，如"各种 println (...) 方法"。如果一组标识符被寻址，则会将其写成 XXX，类似"printlnXXX (...) 会在屏幕上显示 ..."。

为了避免本书内容过于冗余，建议解决方案通常只包含相关的代码片段。文件名会在列表签名中显示，如下所示：

```
class VanillaJava { }
```
列表 1 VanillaJava.java

我们时常需要从命令行调用程序（同义词：命令行、控制台、shell）。由于每个命令行程序都有自己的提示序列，所以在本书中一般用 $ 表示它。用户的条目以粗体设置。例如：

$ java –version
openjdk version "15" 2020-09-15
OpenJDK Runtime Environment (build 15+36-1562)
OpenJDK 64-Bit Server VM (build 15+36-1562, mixed mode, sharing)

如果特别强调是 Windows 命令行，则会使用提示字符">"。例如：

> netstat –e
Schnittstellenstatistik

	Empfangen	Gesendet
Bytes	1755832001	2040354965
Unicastpakete	122394006	119142408
Nicht-Unicastpakete	2120671	163121
Verworfen	0	0
Fehler	0	0
Unbekannte Protok.	0	

船长 CiaoCiao 和 Bonny Brain 的帮助

CiaoCiao 船长和 Bonny Brain 是现代化的海盗，他们在忠诚船员的帮助下在海洋上航行。他们在各大洲做着不正当的生意；土匪和雇佣兵是他们的助手。他们的家乡是巴鲁岛，他们所使用的货币是里雷塔！由于工作人员来自世界各地，所以编程版本也总是用英文书写。

前段时间，CiaoCiao 船长和 Bonny Brain 帮助你解决了问题。现在该还债啦。请帮助他们两个处理他们的事务。这将是值得的！

关于作者

克里斯蒂安·尤伦布姆 10 岁时在 C64 计算机上敲出了他的第一行代码。经过多年的编程和早期的 BASIC 语言扩展，在计算机科学和心理学专业毕业后，他旅行来到了 Java 岛。在 Python、JavaScript、TypeScript 和 Kotlin 的假期旅行还是没能把他从学者症候群中解脱出来。

20 多年来，克里斯蒂安·尤伦布姆一直是一位充满激情的软件架构师、Java 培训师（http://www.tutego.de）和 IT 专家培训师。基于多年的培训经历，他创作了两本著名的专业图书：《Java 岛：程序员经典标准教程》（第 15 版，Rheinwerk 2020，Java 岛 1）和《Java SE 9 标准库：Java 开发人员手册》（第 3 版，Rheinwerk 2017，Java 岛 2）。这些书让读者更接近 Java 编程语言。2005 年，太阳微系统公司（如今的甲骨文公司）表彰了克里斯蒂安·尤伦布姆对 Java 的杰出贡献。太阳微系统公司以及之后的甲骨文公司至今授予了世界各地 300 多人"Java 冠军程序员"的称号。

作为一名 Java 讲师，他不仅会讲解，还会设计各类的练习任务以确保良好的培训效果。在此过程中，他创建了一个庞大并且全面的任务目录，并不断进行扩展和更新。本书汇编了这些任务，包括完整的、记录在案的解决方案。

作为爱好，他在多特蒙德建立了 BINARIUM（http://www.binarium.de），这是最大的家用计算机和游戏机博物馆之一。克里斯蒂安·尤伦布姆生活在莱茵河下游的松斯贝克。

致谢

在此，我要感谢以不同方式为本书的成功做出贡献的每个人。特别感谢校对人员 Michael Rauscher, Thomas Meyer 和 Kevin Schmiechen。

第1章
空间与时间

几乎在所有任务中我们都有屏幕输出，任何在日常生活中用德语写作的人都可能用德语输出。但是，在某些地方总是会出现"错误"的输出，例如：浮点数中的小数位用句点而不是逗号分隔——小数分隔符只是不同国家/地区标准差异的众多示例之一；货币有时放在数字之前，有时放在数字之后；对于日期，有些国家的格式是年–月–日，有些国家的格式是日–月–年，还有些国家的格式是月–日–年。

本章的重点是国际化（如何使程序原则上不受语言影响）和本地化（适应特定语言）的相关任务。如果我们的软件要取得成功，那么它当然必须在任何时间、任何地点都可以运行。Java 可以轻易适应许多语言的特殊性，我们想在任务中了解一下，这样 CiaoCiao 船长和 Bonny Brain 也可以在任何地方做他们的生意，每个人都能理解他们的"语言"。

本章使用的数据类型如下：

- java.util.Locale (https://docs.oracle.com/en/java/javase/11/docs/api/java.base/java/util/Locale.html)
- java.time.LocalDate (https://docs.oracle.com/en/java/javase/11/docs/api/java.base/java/time/LocalDate.html)
- java.time.LocalDateTime (https://docs.oracle.com/en/java/javase/11/docs/api/java.base/java/time/LocalDateTime.html)
- java.time.format.DateTimeFormatter (https://docs.oracle.com/en/java/javase/11/docs/api/java.base/java/time/format/DateTimeFormatter.html)
- java.time.format.FormatStyle (https://docs.oracle.com/en/java/javase/11/docs/api/java.base/java/time/format/FormatStyle.html)
- java.time.format.DateTimeFormatter (https://docs.oracle.com/en/java/javase/11/docs/api/java.base/java/time/format/DateTimeFormatter.html)

- java.time.Duration (https://docs.oracle.com/en/java/javase/11/docs/api/java.base/java/time/Duration.html)

1.1 语言和国家

为了使 Java 库能够解析与格式化浮点数字和日期以及翻译文本，存在数据类型 Locale，它代表一种语言和一个可选的区域。我们想使用这种数据类型来解决一些任务，这些任务很好地显示了 Locale 可以用在哪些地方。

1.1.1 针对随机数应用特定国家／语言的格式化★

Bonny Brain 正在准备一个新的电子邮件骗局：比特币将以远低于售价的价格被"出售"。她为此准备了主题行，如下所示：

Buy Bitcoin for just $11,937.70

当然，船员们正在策划一场全球性的骗局，因此根据不同国家／地区的规则对数字进行格式化很重要。

printf(...) 方法被重载，String.format(...) 也是如此。

- 以 Locale 作为第一个参数。
- 没有 Locale。如果没有传递 Locale 对象，则适用默认 Locale。这会导致 JVM 在德语操作系统中采用德语，在用 System.out.printf(...) 输出时和浮点数时默认使用逗号作为小数分隔符。如果使用英语操作系统，则在默认情况下，句点会被用作分隔符，因为在英语国家小数位是由句点分隔的。

任务：

- 生成一个介于 10 000（包含）和 12 000（不含）之间的 double 类型的随机

数，需要小数位。
- 使用 String.format(String format, Object… args) 方法格式化带两位小数的浮点数，要有千位分隔符。
- 询问系统中的所有 Locale 对象并将它们用作 String.format(Locale l, String format, Object… args) 方法的参数，以便在任何情况下都"本地"格式化浮点数。输出字符串。

> **Java SE API 设计技巧**
> 通常，所有实现与语言相关的格式化或解析字符串的方法都接收一个 Locale 对象作为参数。可能还有其他不带参数的语言相关方法，但这些方法通常是重载的方法，它们在内部请求默认语言，然后转发给带有显式 Locale 参数的方法。

1.2 日期类和时间类

乍一看，日期似乎只包含年、月、日。但是，API 有望回答更多问题：这天是星期三吗？如果一个派对在 2 月 27 日开始，持续 3 天，那么什么时候结束？今年的第 12 周什么时候开始？如果我在 2021 年 12 月 31 日 09:30 离开德国，并在 11 小时后到达迈阿密，那么当地时间是几点？

Java 库多年来不断发展，日期和时间计算也有多种类型：

- java.util.Date（自 Java 1.0 起）；
- java.util.Calendar, java.util.GregorianCalendar, java.util.TimeZone（自 Java 1.1 起）；
- java.time 包（自 Java 8 起，包含类如 LocalDate, LocalTime, LocalDateTime, Duration）。

带有时间成分的日期有三种可能，但是自 Java 8 起 Date 和 Calendar 不再流行，因为它们带来了许多问题。但是，我们仍然可以在许多示例中找到这些数据类型，尤其是在线示例。我们应该远离这些"旧"类型，因此本节专门练习如何使用 java.time 中的当前数据类型。

1.2.1 以不同语言格式化日期输出 ★

12 月 19 日也是海盗话语日（International Talk Like a Pirate Day）。Bonny Brain 正在计划一个聚会并准备邀请函，日期要用 Locale.CHINESE，Locale.ITALIAN 和 new Locale("th") 等语言进行格式化。例如，德国人写 Tag.Monat.Jahr，但其他语言呢？

任务：

- 为 9 月 19 日创建一个 LocalDate 对象：
  LocalDate now = LocalDate.of(Year.now( ).getValue( ), Month.SEPTEMBER, 19);
- 调用 toString( ) 方法，输出是什么？
- 调用 format(DateTimeFormatter.ofLocalizedDate(FormatStyle.MEDIUM) 于 LocalDate，输出是什么？
- 总共有四种 FormatStyle 风格——都试试。使用哪种模式？
- 在 DateTimeFormatter 对象上，我们可以调用 withLocale(Locale) 更改语言。尝试不同的语言。

1.2.2　今年弗朗西斯·博福特爵士的生日是星期几？★

CiaoCiao 船长每年都会庆祝弗朗西斯·博福特爵士的生日，他出生于 1774 年 5 月 27 日。

任务：

- 给出带有弗朗西斯生日的 LocalDate：
  LocalDate beaufortBday = LocalDate.of(1774, Month.MAY, 27);
- 基于 beaufortBday，开发新的带有当前年份的 LocalDate 对象，其中当前年份不是通过硬编码产生的，而是应该从系统中动态产生的。
- 创建一个输出，显示今年弗朗西斯·博福特爵士在星期几庆祝他的生日（以何种形式输出星期几，即数字或字符串或哪种语言都可以）。

举例：

2020 年的输出如下：

WEDNESDAY
3
Mittwoch

1.2.3　确定卡拉 OK 派对的平均时长★

有朗姆酒、跳舞和唱歌的卡拉 OK 派对很受船员欢迎。卡拉 OK 派对常常持续到黎明，这让 Bonny Brain 很烦恼，因为船员们第二天就会打瞌睡。

为了算出这些过度行为平均持续了多少小时，Bonny Brain 想绘制统计数据。她写下了开始和结束的时间，之后可以计算出平均数。例如，在纸条上写"2022-03-12, 20:20 – 2022-03-12, 23:50"。

任务：

- 编写一个程序，提取上述格式的字符串，确定并输出卡拉OK派对的平均持续时间。
- 该程序不必考虑时区、闰秒或其他特殊情况——一天的时间正好是24小时。

举例：

字符串如下：

```
2022-03-12, 20:20 - 2022-03-12, 23:50
2022-04-01, 21:30 - 2022-04-02, 01:20
```

输出如下：

```
3 h 40 m
```

> **小提示：**
> Duration 类有助于解决时差问题。

1.2.4 解析不同的日期格式 ★★★

一个日期可以被指定为绝对的或相对的，并且有几种方法来指定日期。举几个例子：

```
2020-10-10
2020-12-2
1/3/1976
1/3/20
tomorrow
today
yesterday
1 day ago
2234 days ago
```

任务：

- 编写一个 Optional<LocalDate> parseDate(String string) 方法，它可以识别上面提到的格式。
- 如果字符串是其中的一种格式，则该方法应解析该字符串，将其转换为 LocalDate 并以 Optional 返回。
- 如果无法解析出任何格式，则返回 Optional.empty( )。

1.3 可供参考的解决方案

任务 1.1.1：针对随机数应用特定国家 / 语言的格式化

```
double random = ThreadLocalRandom.current().nextDouble( 10_000, 12_000 );
Locale[] locales = Locale.getAvailableLocales();
for ( Locale locale : locales )
   System.out.printf( locale, "%,.2f (%s)%n", random, locale.getDisplayName() );
```

列表 1.1 com/tutego/exercise/util/RandomInEveryLocalePrinter.java

解决方案包括四个部分：第一部分，生成一个随机数；第二部分，将所有注册的 Locale 对象作为一个数组获取，数组包含 Java 库所有支持的语言；第三部分，遍历这个数组；第四部分，输出。

System.out.printf(...) 需要一个格式化字符串，其中包含作为浮点数的随机数，就像语言名称一样。字符串的数字格式是 %,.2f——逗号表示需要千位分隔符，".2" 代表两位小数。

printf(...) 中的重要参数是第一个，它表示可以传递一个 Locale 实例，在下文中它决定了浮点数的格式化。

任务 1.2.1：以不同语言格式化日期输出

```
LocalDate date = LocalDate.of( Year.now().getValue(), Month.SEPTEMBER, 19 );
System.out.println( date );

DateTimeFormatter formatterShort =
  DateTimeFormatter.ofLocalizedDate( FormatStyle.SHORT );
DateTimeFormatter formatterMedium =
  DateTimeFormatter.ofLocalizedDate( FormatStyle.MEDIUM );
DateTimeFormatter formatterLong =
  DateTimeFormatter.ofLocalizedDate( FormatStyle.LONG );
DateTimeFormatter formatterFull =
  DateTimeFormatter.ofLocalizedDate( FormatStyle.FULL );

System.out.println( date.format( formatterShort ) );
System.out.println( date.format( formatterMedium ) );
System.out.println( date.format( formatterLong ) );
System.out.println( date.format( formatterFull ) );

System.out.println( date.format( formatterShort.withLocale(
  Locale.CANADA_FRENCH ) ) );
System.out.println( date.format( formatterMedium.withLocale(
  Locale.CHINESE ) ) );
System.out.println( date.format( formatterLong.withLocale(
  Locale.ITALIAN ) ) );
System.out.println( date.format( formatterFull.withLocale(
  new Locale( "th" ) ) ) );
```

列表 1.2 com/tutego/exercise/time/DateTimeFormatterDemo.java

程序针对 2021 年的输出如下：

```
021-09-19
19.09.21
19.09.2021
19. September 2021
Sonntag, 19. September 2021
2021-09-19
```

2021 年 9 月 19 日
19 settembre 2021
วันอาทิตย์ที่ 19 กันยายน ค.ศ. 2021

时间数据类型重写 toString(...) 方法，然而，该方法不能用格式化类型进行参数化。相反，LocalDate 有 format(...) 方法，它可以传递一个 DateTimeFormatter。获取 DateTimeFormatter 的常见方法有三种：

- 选择一个常量，如 DateTimeFormatter.ISO_LOCAL_DATE。
- 使用模式进行精准确定，例如日或月位于哪里，以及使用了哪些分隔符。
- 选择独立于语言的标准格式，这对屏幕输出非常实用。它提供 DateTimeFormatter.ofLocalizedDate(...)。

ofLocalizedDate(...) 方法希望得到一个 FormatStyle 类型的参数。建议解决方案建立了四个这样的 DateTimeFormatter 实例，并用这些格式调用 format(...) 方法。

方法名 LocalizedDate(...) 已经包含了一个对本地化的引用。任何 DateTimeFormatter 都可以用 WithLocale(...) 与一个 Locale 关联。所有时间数据类型都是不可改变的，方法返回的是一个带有 Locale 的新对象。

任务 1.2.2：今年弗朗西斯·博福特爵士的生日是星期几？
分两步来解决这个问题：

1. 从 beaufortBday 开始，必须建立一个以今年 5 月 27 日为日期的 LocalDate。
2. 必须询问今年弗朗西斯·博福特爵士的生日是星期几。

第一步：
```
LocalDate beaufortBday = LocalDate.of( 1774, Month.MAY, 27 );

// 1.
LocalDate beaufortBdayThisYear = beaufortBday.withYear( Year.now().getValue() );

// 2.
LocalDate beaufortBdayThisYear2 = LocalDate.of( LocalDate.now().getYear(),
    beaufortBday.getMonth(),
    beaufortBday.getDayOfMonth() );
```

```
// 3.
LocalDate beaufortBdayThisYear3 = LocalDate.now()
    .withMonth( beaufortBday.getMonthValue() )
    .withDayOfMonth( beaufortBday.getDayOfMonth() );
```
列表 1.3 com/tutego/exercise/time/SirFrancisBeaufortBirthday.java

建议解决方案显示了创建 LocalDate 对象的三种变体：

1. 第一个变体，我们使用了一个 Wither，即一个带有前缀 with（不是 set）的方法，它返回一个具有修改值的新对象。Java 的时间数据类型是不可变的，因此没有 Setter。withYear(int) 返回一个带有修改年份的新的 LocalDate 对象。对于当前年份，我们可以使用数据类型 Year。静态方法 now( ) 返回当前年份，由于 withYear(int) 需要一个整数，所以 Year 对象中的 getValue( ) 是必要的。
2. 第二个变体，我们可以使用静态工厂方法 of(...) 来构造一个新的 LocalDate 对象，而不使用 Wither，方法是获取当前年份并询问 beaufortBday 的月份和日期。
3. 第三个变体，我们使用 LocalDate.now( ) 查询当前日期的日、月、年，但是使用两个 Wither 方法，首先获取一个带有设置月份的新的 LocalDate 对象，然后是一个带有该设置月份的日期的新的 LocalDate 对象。这个变体并不是最优的，因为会创建两个临时的 LocalDate 对象，它们接着会进入自动垃圾回收阶段。

第二步：
```
DayOfWeek dayOfWeek = beaufortBdayThisYear.getDayOfWeek();
System.out.println( dayOfWeek );
System.out.println( dayOfWeek.getValue() );
System.out.println( dayOfWeek.getDisplayName( TextStyle.FULL,
    Locale.GERMANY ) );

DateTimeFormatter formatter =
    DateTimeFormatter.ofPattern( "EEEE" /*, Locale.GERMANY */ );
System.out.println( beaufortBdayThisYear.format( formatter ) );
```
列表 1.4 com/tutego/exercise/time/SirFrancisBeaufortBirthday.java

建议解决方案显示了输出星期几的不同做法。
1. LocalDate 对象提供各种 Getter。getDayOfWeek( ) 返回枚举 DayOfWeek，我们得到星期几。DayOfWeek 是一个枚举类型，由于所有枚举都有一个 toString( )

方法，所以输出如 WEDNESDAY。如果需要一个数字值，则 DayOfWeek 的 getValue( ) 会返回一个数字，如 3 代表星期三。

2. DayOfWeek 提供了实用的方法 getDisplayName(TextStyle style, Locale locale)，它返回一个用当地语言格式化的星期日期，例如德语对应的字符串为 "Mittwoch"。

3. 另一个变体不通过 DayOfWeek 工作，而是使用 LocalDate 提供的 format(..) 方法。这需要传递一个 DateTimeFormatter，如果我们使用 EEEE 模式，那么它代表星期日期。可以选择性地指定星期日期的语言，如果缺少该语言，则以操作系统的默认语言为准。

任务 1.2.3：确定卡拉 OK 派对的平均时长

```
String input = "2022-03-12, 20:20 - 2022-03-12, 23:50\n" +
               "2022-04-01, 21:30 - 2022-04-02, 01:20";

DateTimeFormatter formatter = DateTimeFormatter.ofPattern(
    "yyyy-MM-dd, HH:mm" );
Scanner scanner = new Scanner( input ).useDelimiter( " - |\\n" );
Duration totalDuration = Duration.ZERO;

int lines;
for ( lines = 0; scanner.hasNext(); lines++ ) {
  String start = scanner.next();
  String end = scanner.next(); // potential NoSuchElementException

  // potential DateTimeParseException
  LocalDateTime startDateTime = LocalDateTime.parse( start, formatter );
  LocalDateTime endDateTime = LocalDateTime.parse( end, formatter );

  Duration duration = Duration.between( startDateTime, endDateTime );
  totalDuration = totalDuration.plus( duration );
}

Duration averageDuration = totalDuration.dividedBy( lines );
System.out.printf( "%d h%02d m", averageDuration.toHours(),
    averageDuration.toMinutesPart() );
```

列表 1.5 com/tutego/exercise/time/AverageDuration.java

完成这个任务需要多个步骤。首先，程序必须从大字符串中提取所有的开始值和结束值。然后，这些字符串必须被转移到相应的时间数据类型。接着，必须计算这些开始时间和结束时间的差，并将其相加。最后，把总和除以条目数，任务就完成了。

Scanner 可以接管对开始时间和结束时间的识别。在默认情况下，Scanner 初始化时使用空白作为分隔符。我们改变这一点，设置一个减号和一个换行符作为分隔符。重复调用 next() 方法，连续返回开始和结束的值。

提取字符串后，我们将其传递给 LocalDateTime 类的 parse(...) 方法。由于我们的字符串不是基于 ISO 标准的，所以必须通过 parse(...) 传递一个 DateTimeFormatter。我们预先使用 ofPattern(...) 方法创建它，其中传递的字符串与我们用来格式化日期和时间的模式完全对应。符号序列 YYYY，MM，dd 等可以从 Java 文档中获取。

使用两个 LocalDateTime 对象，可以确定差——我们不手动执行此操作，而是使用 Duration 类。Java-Date-Time API 中有两个类用于时间数据类型的差：Duration 类以秒为单位存储两个时间值之间的间隔，而 Period 用于日期值并以天为单位工作。如果差值中包含闰日，那么使用 Period 很合适；如果使用 Duration，那么一天正好是 24 小时，相当于 86 400 秒。Duration 适用于我们的计算。

直接用 between(...) 方法建立一个 Duration 对象传递开始值和结束值很实用。与所有时间数据类型一样，Duration 是不可变的。为了计算差的和，我们使用 plus(...) 方法并将结果存储回 totalDuration 变量。最后，增加一个变量，即日期-时间对的总数，然后回到循环测试，看是否还有其他日期。

程序中可能出现两个运行时错误。因为字符串中的日期数量是奇数，当然，也因为日期时间的格式化是错误的，所以我们不会捕获这些异常，它们会导致程序终止。

如果所有的值都是正确的，程序就以计算平均持续时间结束。dividedBy(...) 方法有助于将总和的持续时间除以出现的日期-时间对的数量，然后向 Duration 类型产生的对象请求小时数和分钟数，并对整体进行格式化。

任务 1.2.4：解析不同的日期格式

这项任务的挑战在于每个日期的格式并不统一，而我们有不同的格式。Java 库实际上通过 DateTimeFormatterBuilder 提供对多种不同格式的支持。通过 appendPattern(...) 方法，可以指定一个模式来确定格式。把格式放在方括号中，表示它是可选的。例如：

```
DateTimeFormatter formatter = new DateTimeFormatterBuilder()
    .appendPattern( "[d/M/yyyy]" )
    .appendPattern( "[yyyy-MM-dd]" )
```

```
    .parseDefaulting( ChronoField.MONTH_OF_YEAR,
        YearMonth.now().getMonthValue() )
    .parseDefaulting( ChronoField.DAY_OF_MONTH,
        LocalDate.now().getDayOfMonth() )
    .toFormatter();
```

在我们的例子中，不能使用以这种方式配置的 DateTimeFormatter，因为无法表达"昨天""今天"或"明天"等相对表达。原则上，可以使用以这种方式配置的 DateTimeFormatter 来实现一部分，但建议解决方案采用了其他方法。

```
public static Optional<LocalDate> parseDate( String string ) {
  LocalDate now = LocalDate.now();
  Collection<Function<String, LocalDate>> parsers = Arrays.asList(
      input -> LocalDate.parse( input, DateTimeFormatter.ofPattern(
          "yyyy-M-d" ) ),
      input -> LocalDate.parse( input, DateTimeFormatter.ofPattern(
          "d/M/yyyy" ) ),
      input -> LocalDate.parse( input, DateTimeFormatter.ofPattern(
          "d/M/yy" ) ),
      input -> input.equalsIgnoreCase( "yesterday" ) ?
          now.minusDays( 1 ) : null,
      input -> input.equalsIgnoreCase( "today" ) ? now : null,
      input -> input.equalsIgnoreCase( "tomorrow" ) ?
          now.plusDays( 1 ) : null,
      input -> new Scanner( input ).findAll( "(\\d+) days? ago" )
          .map( matchResult -> matchResult.group( 1 ) )
          .mapToInt( Integer::parseInt )
          .mapToObj( now::minusDays )
          .findFirst().orElse( null ) );

  for ( Function<String, LocalDate> parser : parsers ) {
    try { return Optional.of( parser.apply( string ) ); }
    catch ( Exception e ) { /* Ignore */ }
  }
  return Optional.empty();
}
```

列表 1.6 com/tutego/exercise/time/ParseDatePattern.java

最后，我们的任务是将 String 转换为 LocalDate，但这只不过是一个映射，我们可以在 Java 中使用数据类型 Function 来表达它。可以编写一个函数来尝试识别某种格式。最好的情况是映射有效，可以将字符串映射到 LocalDate；最坏的情况是出现异常，或者将此函数编程为返回 null。这种概括乍一看似乎很不寻常，其原因在于我们之后可以收集所有 Function 对象，并一个接一个地进行尝试。在编程时，我们始终应该考虑是否可以概括事物——从具体到一般。

我们可以识别三种不同类型的映射：

- 前三个映射试图使用 DateTimeFormatter 直接识别一个模式。
- 第二种类型的函数识别今天、昨天和明天。我们使用本地变量 now，在进入方法时初始化。其用于测试时很笨拙，因此在实践中，我们将为测试用例编写一个内部方法，可以在其中引入 LocalDate 作为参考点。
- 第三个函数是最复杂的，因为它需要识别给定天数的相对关系。这里有几种方法。Scanner 可以识别字符串并且可以提供天数的查找。结果在 MatchResult 中：我们提取天数，将其转换为整数，获取新的 LocalDate 对象，对天数进行换算，然后返回结果。如果模式不匹配，则将返回 null。

现在每种可能的格式都有一个 Function，如果要识别更多的格式，则还要增加新的映射，这是一项艰苦的工作。实际识别是通过遍历所有映射来进行的；这些函数依次应用于字符串，直到 Function 返回有效结果。为此，函数首先包含在列表中。如果应用没有工作并且出现异常，则 catch 块会捕获异常并从列表中选择下一个函数。相对引用都以 null 回应，Optional.of(...) 会导致 NullPointerException，该异常被捕获以便可以使用下一个函数。

最后只有两种退出：要么函数以有效的 LocalDate 响应，我们用 return 退出 try 块；要么只有异常，for 循环找不到任何候选者——方法以 Optional.empty() 退出。

第 2 章
基于线程的并发编程

操作系统提供线程作为一种编程辅助工具，程序一直勤于利用线程。任何使用 Windows 工作的人，只要看一眼任务管理器，就会发现有 3 000 个线程在快速运行。在本章中，我们再次增大这个数字，以便让多个活动并发甚至并行运行。这些任务不应该只是创建线程，如果某些东西并发运行，那么还必须确保对公共资源的正确访问——线程必须相互协调。

本章使用的数据类型如下：

- java.lang.Thread (https://docs.oracle.com/en/java/javase/11/docs/api/java.base/java/lang/Thread.html)
- java.lang.Runnable (https://docs.oracle.com/en/java/javase/11/docs/api/java.base/java/lang/Runnable.html)
- java.util.concurrent.TimeUnit (https://docs.oracle.com/en/java/javase/11/docs/api/java.base/java/util/concurrent/TimeUnit.html)
- java.util.concurrent.Callable (https://docs.oracle.com/en/java/javase/11/docs/api/java.base/java/util/concurrent/Callable.html)
- java.util.concurrent.Executor (https://docs.oracle.com/en/java/javase/11/docs/api/java.base/java/util/concurrent/Executor.html)
- java.util.concurrent.Executors (https://docs.oracle.com/en/java/javase/11/docs/api/java.base/java/util/concurrent/Executors.html)
- java.util.concurrent.Future (https://docs.oracle.com/en/java/javase/11/docs/api/java.base/java/util/concurrent/Future.html)
- java.util.concurrent.locks.Lock (https://docs.oracle.com/en/java/javase/11/docs/api/java.base/java/util/concurrent/locks/Lock.html)
- java.util.concurrent.lock.ReentrantLock (https://docs.oracle.com/en/java/javase/11/docs/api/java.base/java/util/concurrent/locks/Lock.html)

- java.util.concurrent.Semaphore (https://docs.oracle.com/en/java/javase/11/docs/api/java.base/java/util/concurrent/Semaphore.html)
- java.util.concurrent.locks.Condition (https://docs.oracle.com/en/java/javase/11/docs/api/java.base/java/util/concurrent/locks/Condition.html)
- java.util.concurrent.CyclicBarrier (https://docs.oracle.com/en/java/javase/11/docs/api/java.base/java/util/concurrent/CyclicBarrier.html)
- java.util.concurrent.CountDownLatch (https://docs.oracle.com/en/java/javase/11/docs/api/java.base/java/util/concurrent/CountDownLatch.html)

2.1 创建线程

当 JVM 启动时，它会生成一个名为 main 的线程。该线程运行 main(...) 方法，并在所有之前的任务中运行了我们的程序。我们想在接下来的任务中改变这一点。我们想创建更多线程并让它们处理程序代码（图 2.1）。

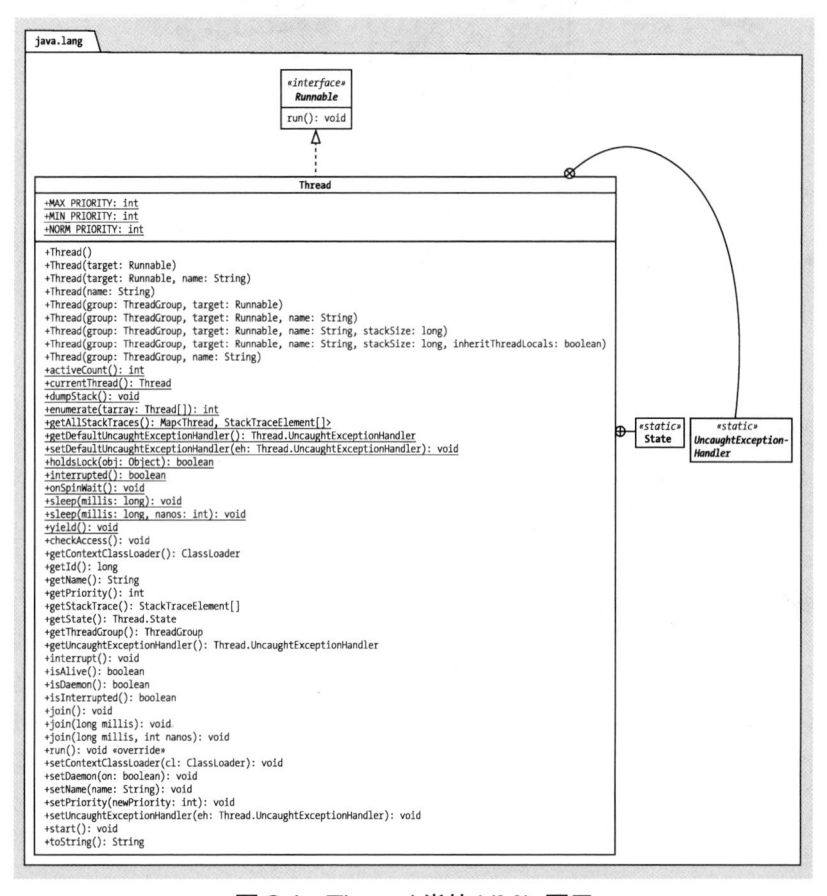

图 2.1 Thread 类的 UML 图示

2.1.1 创建用于招手和挥动旗帜的线程 ★

为表示对 CiaoCiao 船长的崇敬之情，人们举行了一场游行。CiaoCiao 船长站在船舷梯上，一只手在打招呼，另一只手挥舞着一面旗帜。

任务：

- 在 Java 中，线程总是执行 Runnable 类型的东西。Runnable 是一个函数式接口，在 Java 中有两种方法来实现函数式接口：类和 Lambda 表达式。编写两个 Runnable 的实现，一个使用类，一个使用 Lambda 表达式。
- 在两个 Runnable 的实现中，设置一个重复 50 次的循环。一个 Runnable 应该在屏幕上输出 "winken"（招手），另一个 Runnable 应该在屏幕上输出 "Fähnchen schwenken"（挥动旗帜）。
- 创建一个 Thread 对象，并传递 Runnable，然后启动线程。不要启动 50 个线程，而只启动 2 个！

拓展：

每个线程的 run( ) 方法应该包含 "System.out.println(Thread.currentThread( ));" 这一行。这样会显示什么？

假设 CiaoCiao 船长还有几只手臂可以挥舞。在系统"静止"之前，可以创建多少个线程？观察 Windows 任务管理器（按"Ctrl+Alt+Del"组合键）中的内存消耗情况。估计一条线程的"成本"是多少。

2.1.2 不再招手或挥动旗帜：结束线程 ★

线程可以用 stop( ) 硬性终止——该方法已被废弃了几十年，但可能永远不会被删除——或者用 interrupt( ) 请求自行终止。然而，要做到这一点，线程必须配合并使用 isInterrupted( ) 来检查是否存在这样的删除请求。

CiaoCiao 船长仍然站在船上，招手并挥舞着旗帜。回归正常，他现在必须停止这种娱乐活动。

任务：

- 编写一个有两个 Runnable 实现的程序，原则上除非有中断，否则 CiaoCiao 船长会无止境地招手和挥动旗帜。因此，run( ) 方法应该通过 Thread.currentThread( ).isInterrupted( ) 测试是否有中断，然后退出循环。
- 在循环中建立一个延迟。复制以下代码：
  ```
  try { Thread.sleep (2000); } catch ( InterruptedException e ) {
  Thread.currentThread().interrupt(); }
  ```
- 主程序应使用 JOptionPane.showInputDialog(String) 对输入做出反应，以便使命令 endw 结束招手，命令 endf 结束挥动旗帜。

2.1.3 参数化 Runnable ★★

从下面的代码可以看出，两个 Runnable 的实现非常相似，只是屏幕输出不同（run( ) 方法的不同用粗体表示）：

```
// Runnable 1
class Wink implements Runnable {
  @Override public void run() {
    for ( int i = 0;i< 50; i++ )
      System.out.printf( "Wink;%s%n", Thread.currentThread() );
  }
}
Runnable winker = new Wink();

// Runnable 2
Runnable flagWaver = () -> {
  for ( int i = 0;i< 50; i++ )
    System.out.printf( "Wave flag;%s%n", Thread.currentThread() );
};
```

但是，代码重复并不是好事，应该改变这一点。

任务：
- 思考：原则上如何为 Runnable 添加一些东西？
- 实现一个参数化 Runnable，以便在上面的循环中：
  - 可以自由确定屏幕输出；
  - 重复次数。
- 重写招手和挥动旗帜的程序，以便将参数化的 Runnable 传递给线程执行。

2.2 执行和休眠

一个线程可以处于多种状态，包括运行、等待、睡眠、阻塞或终止。在前面的任务中，我们启动了线程，使其处于运行状态，并通过结束循环来终止 run() 方法，这也终止了线程。在这一节中，任务是关于睡眠状态的，而在第 2.4 节 "保护临界区" 和第 2.5 节 "线程协合作和同步助手" 中，有关于等待/阻塞状态的任务。

2.2.1 使用睡眠线程延迟处理 ★★

UNIX 中已知的程序 sleep（睡眠）(https://man7.org/linux/man-pages/man1/sleep.1.html) 可以从命令行调用，然后休眠一段时间，从而延迟脚本中接下来的程序。

任务：
- 在 Java 中重新实现 sleep 程序，以便可以在类似示例的命令行中编写：
  $ java Sleep 22
  然后 Java 程序应该休眠 22 秒，假如脚本中有后续程序调用，则其将被延迟。
- Java 程序应该能够在命令行中以各种方式指定休眠时间。如果只传递一个整数，则等待时间为秒。应允许在整数后添加不同持续时间的后缀：
  - s 表示秒（默认）；
  - m 表示分钟；
  - h 表示小时；
  - d 表示天。
  如果传递了多个值，则将它们相加并得出总等待时间。
- 调用可能出现各种问题，例如，没有传递数字或数字太大。检查值、范围和后缀是否正确。可选：如果发生错误，通过 System.exit(int) 以独特的退出代码结束程序。

举例：
- 有效的调用示例：
  $ java Sleep 1m
  $ java Sleep 1m 2s
  $ java Sleep 1h 3h 999999s
- 无效的、导致中断的调用：
  $ java Sleep
  $ java Sleep three
  $ java Sleep 1y
  $ java Sleep 9999999999999999999999

> **小提示：**
> **构建程序，使三个主要部分是可识别的：**
> 1. 运行命令行并分析传输情况；
> 2. 将单位转换为秒；
> 3. 实际睡眠时长为累计秒数。

2.2.2 通过线程观察文件变化★

一次成功的掠夺后，所有新宝藏都会被系统地添加到库存列表中并被保存在一个简单的文件中。Bonny Brain 希望在文件发生更改时收到通知。

任务：
- 使用构造函数 FileChangeWatcher(String filename) 编写类 FileChangeWatcher。
- 实现 Runnable。
- 每半秒输出一次文件名、文件大小和 Files.getLastModifiedTime(path)。
- 用 getLastModifiedTime(…) 检查文件是否发生变化。如果文件发生变化，则发出消息。所有这些都应该无休止地发生，每个新的变化都应该被报告。
- 拓展：我们现在希望对变化做出更灵活的反应。为此，应该能够在构造函数中传递 java.util.function.Consumer 对象。FileChangeWatcher 应该记住消费者并在发生变化时调用 accept(Path) 方法。我们可以注册一个对象，当文件更改时将通知该对象。

2.2.3 捕捉异常★

已检查的异常和未检查的异常之间的区别有一个重要的后果，因为未检查的异常如果没有被捕获，可能不断恶化并最终出现在执行线程上，从而该执行线程由 JVM 终止。运行时环境会自动执行此操作，我们在标准错误通道上会收到一条消息，但无法恢复线程。

UncaughtExceptionHandler 既可以安装在本地线程上，也可以安装在所有线程上，如果线程因异常终止，则会收到通知。它可以用于四种场景：

1. UncaughtExceptionHandler 可以在单个线程上设置。每当这个线程得到一个未处理的异常时，线程就会被中止并通知设置的 UncaughtExceptionHandler。
2. UncaughtExceptionHandler 可以被设置在线程组上。
3. UncaughtExceptionHandler 可以被全局设置在所有线程上。
4. main 线程的特殊之处在于 JVM 会自动创建它并运行主程序。当然，在 main 线程中也可能存在未检查的异常，UncaughtExceptionHandler 可以报告它。但是，有一个有趣的特殊之处：throws 可以出现在 main(...) 方法中，因此已检查的异常可以返回到 JVM。在异常已检查时，还会通知设置好的 UncaughtExceptionHandler。

级联中的处理：如果存在未检查的异常，则 JVM 首先检查是否在单个线程上设置了 UncaughtExceptionHandler。如果没有，则它会在线程组上查找 UncaughtExceptionHandler，然后查找收到通知的全局处理程序。

任务：
- 运行一个通过除以 0 终止的线程。使用全局 UncaughtExceptionHandler 记录此异常。
- 启动第二个线程，该线程具有忽略异常的本地 UncaughtExceptionHandler，因此也不会出现任何消息。
- 如果 throws Exception 在 main 方法上，并且主体包含一个 new URL("captain")，那么全局的 UncaughtExceptionHandler 是否被调用？

2.3 线程池和结果

对于 Java 开发者来说，自己创建线程并将其连接到程序代码上并不总是最好的方法；将程序代码与物理执行分开往往是更明智的做法。在 Java 中，这由 Executor 完成。它使得将程序代码与实际线程分开成为可能，也可以将同一线程多次用于不同的程序代码。

在 Java 库中有三个中央 Executor 的实现：ThreadPoolExecutor，Scheduled-ThreadPoolExecutor 和 ForkJoinPool。ThreadPoolExecutor 和 ForkJoinPool 类型实现了管理现有线程集合的线程池，这样任务就可以被传递给现有的、空闲的线程。

后台的每个代码执行都是通过 Java 中的一个线程实现的，这个线程要么是自己生成并启动，要么由 Executor 或内部线程池间接启动。有两个重要的接口来封装并发代码：Runnable 和 Callable。Runnable 被直接传递给线程构造函数，Callable 不能被传递给线程。对于 Callable，我们需要一个 Executor。Callable 也会返回一个结果，就像 Supplier 一样，但它没有参数可以传递。Runnable 无任何返回，也不能传递。run( ) 方法不抛出异常，call( ) 在方法签名中有 throws Exception，因此它可以转发任何异常（见图 2.2）。

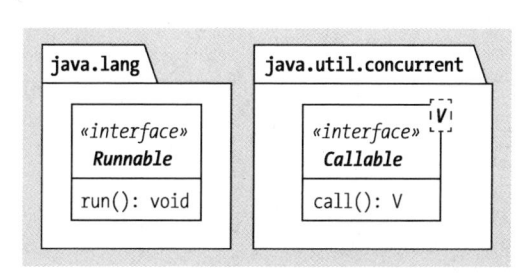

图 2.2　Runnable 和 Callable 接口的 UML 图示

到目前为止，我们一直都是自己构建线程并且只使用 Runnable。接下来的任务将处理线程池以及 Callable。

2.3.1　使用线程池★★

复活节快到了，Bonny Brain 和她的船员装扮成伍基人去孤儿院分发礼物。

任务：

- 使用 Executors.newCachedThreadPool( ) 创建一个 ExecutorService，这是线程池。
- 创建一个礼品字符串数组。
- Bonny Brain 在 main 线程中运行每个礼物，并以 1~2 秒的间隔将其传输给船员，即线程池中的一个线程。
- 船员是线程池中的线程。他们执行 Bonny Brain 的命令，送出一份礼物。为此，他们需要 1~4 秒。
- 过程如下：礼物分配是由一个 Runnable，即原本的动作来实现的。选择线程池中的一个空闲线程（船员）并运行 Runnable。Runnable 需要一种方式来接收 Bonny Brain 的礼物。

2.3.2　确定网站的最后修改★★

下面的类实现了一个方法，该方法返回最后一次修改网页时的时间戳（数据可能不可用，在这种情况下，时间是 1/1/1970）。服务器应在区域时间 UTC ± 0 中传输。

```java
import java.io.IOException;
import java.net.HttpURLConnection;
import java.net.URL;
import java.time.Instant;
import java.time.ZoneId;
import java.time.ZonedDateTime;

public class WebChecker {

  public static void main(String[] args) throws IOException {
    ZonedDateTime urlLastModified = getLastModified(new URL(
        "http://www.tutego.de/index.html"));
    System.out.println(urlLastModified);
    ZonedDateTime urlLastModified2 = getLastModified(new URL(
        "https://en.wikipedia.org/wiki/Main_Page"));
    System.out.println(urlLastModified2);
  }

  private static ZonedDateTime getLastModified(URL url) {
    try {
      HttpURLConnection con = (HttpURLConnection) url.openConnection();
      long dateTime = con.getLastModified();
      con.disconnect();
      return ZonedDateTime.ofInstant(
          Instant.ofEpochMilli( dateTime ), ZoneId.of( "UTC" ) );
    } catch ( IOException e ) {
      throw new IllegalStateException(e);
    }
  }
}
```

任务：

- 创建新类 WebResourceLastModifiedCallable。
- 给 WebResourceLastModifiedCallable 一个构造函数，这样我们就可以传入一个 URL。
- 让 WebResourceLastModifiedCallable 实现 Callable<ZonedDateTime> 接口。将

示例中的 getLastModified(URL) 的实现放在 call( ) 方法中。call( ) 是否必须自己捕获已检查的异常?
- ▶ 创建 WebResourceLastModifiedCallable 对象并让线程池执行它们。
  - 让 Callable 运行一次,没有时间限制。
  - 如果只给 Callable 1 微秒的执行时间,那么结果是多少?
- ▶ 可选:计算网站更改距离当前时间多少分钟。

2.4 保护临界区

如果多个程序部分同时运行,则它们可能访问公共资源或内存区域。这种访问必须是同步的,以便一个线程可以完成工作,直到另一个线程访问该资源。如果对公共资源的访问没有得到协调,就会出现故障状态。

程序必须保护临界区,以便只有一个其他线程可以驻留在一个部分中。Java 提供了两种机制:

1. 关键词 synchronized;
2. Lock 对象。

synchronized 是一个实用的关键词,但性能有限。Java 并发工具包提供了更强大的数据类型。用于"锁定"独占执行的程序部分存在接口 java.util.concurrent.locks.Lock 及其多种实现,如 ReentrantLock、ReentrantReadWriteLock.ReadLock、ReentrantReadWriteLock.WriteLock(见图 2.3)。

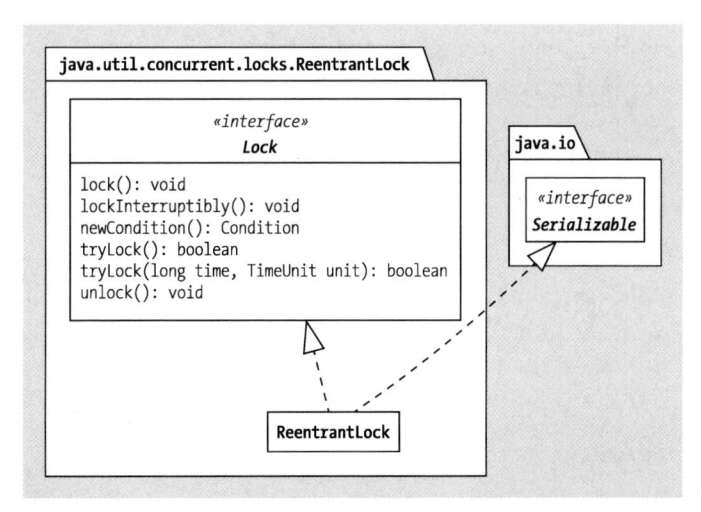

图 2.3 Lock 和 ReentrantLock 类型的 UML 图示

2.4.1 在诗集里写下回忆★

在一艘载有珍贵猪笼草的货船成功地更换了"主人"之后，海盗们在一本诗集中写下了他们的回忆，CiaoCiao 船长后来用贴纸装饰了这本诗集。

以下程序代码在一个类的 main(...) 方法中给出：

```
class FriendshipBook {
  private final StringBuilder text = new StringBuilder();
  public void appendChar( char character ) {
    text.append( character );
  }

  public void appendDivider() {
    text.append(
        "\n_,.-´~´-.,__,.-´~´-.,__,.-´~´-.,__,.-´~´-.,__,.-´~´-.,_\n"
);
  }

  @Override public String toString() {
    return text.toString();
  }
```

```java
}

class Autor implements Runnable {
  private final String text;
  private final FriendshipBook book;

  public Autor( String text, FriendshipBook book ) {
    this.text = text;
    this.book = book;
  }

  @Override public void run() {
    for ( int i = 0; i < text.length(); i++ ) {
      book.appendChar( text.charAt( i ) );
      try { Thread.sleep( 1 ); }
      catch ( InterruptedException e ) { /* Ignore */ }
    }
    book.appendDivider();
  }
}

FriendshipBook book = new FriendshipBook();

String q1 = "Die Blumen brauchen Sonnenschein " +
    "und ich brauch Capatain CiaoCiao zum Fröhlichsein";
new Thread( new Autor( q1, book ) ).start();

String q2 = "Wenn du lachst, lachen sie alle. " +
    "Wenn du weinst, weinst du alleine";
new Thread( new Autor( q2, book ) ).start();

TimeUnit.SECONDS.sleep( 1 );

System.out.println( book );
```

任务：
- 在运行程序之前，考虑预期的结果。
- 把代码放在自己的类和 main(...) 方法中，并检查假设。
- 程序的缺点是对 FriendshipBook 无限制访问。使用 Lock 对象改进程序，使 FriendshipBook 一次只能由一个海盗写入。

2.5 线程协作和同步助手

同步程序代码是很重要的，这样两个线程就不会互相覆盖对方的数据。我们已经看到，这对 Lock 对象有效。然而，Lock 对象只锁定一个关键区域，当一个关键区域被锁定时，运行环境会自动让一个线程等待。这个机制的扩展是让一个线程或多个线程不仅等待进入关键区域的许可，而且还能通过信号被通知执行相应任务。Java 提供了不同的同步辅助工具，它们有一个内部状态，当满足某些条件时，可以让其他线程等待或开始运行。

- Semaphore：虽然 Lock 只允许一个线程驻留在一个临界区中，但 Semaphore 允许用户定义数量的线程驻留在块中。方法名称也略有不同：Lock 声明 lock( ) 方法，而 Semaphore 声明 acquire( ) 方法。如果使用 acquire( ) 达到最大数量，则线程必须像 Lock 一样等待访问。最大计数为 1 的 Semaphore 就像 Lock。
- Condition：通过 Condition，线程可以把自己搁置，然后被另一个线程再次唤醒。Condition 对象可用于编排消费者-生产者关系，但在实践中很少需要这种数据类型，因为有基于它的 Java 类型，其通常更简单、更灵活。Condition 是一个接口，Lock 对象提供工厂方法来提供 Condition 实例。
- CountDownLatch：CountDownLatch 类型的对象被初始化为一个整数，各线程对这个 CountDownLatch 进行倒数，使其处于等待状态。当 CountDownLatch 最终达到 0 时，所有线程又被释放。因此，CountDownLatch 是一种将不同的线程集中在一个共同点上的方法。一旦一个 CountDownLatch 被用完，它就不能被重置。
- CyclicBarrier：该类是所谓的屏障的实现。屏障允许多个线程在一个点相遇。如果工作订单是并行处理的，并且必须稍后再次放在一起，则可以使用屏障来实现。在所有线程在此屏障处聚集在一起之后，它们会继续运行。CyclicBarrier 的构造函数可以传递一个 Runnable，以便重合时调用。与 CountDownLatch 不同，CyclicBarrier 可以重置和再次使用。
- Exchanger：生产者-消费者关系在程序中很常见，生产者将数据传输给消费者。正是在这种情况下使用类 Exchanger。它允许两个线程在一点相遇并交换数据。

2.5.1 与船长一起参加宴会——Semaphore ★★

Bonny Brain 和 CiaoCiao 船长计划与同伴一起举办宴会。他们都坐在一张有 6 个座位的桌子旁，接待不同的客人。客人们来了，稍作停留、谈笑风生、吃些东西，然后又离开他们。

任务：
- 创建一个 Semaphore，其座位数与同时坐在餐桌上的客人数量一样多。
- 将客人建模为一个实现 Runnable 的 Guest 类。所有客人都有一个名称。
- 客人们都在等着入座。让等待时间最长的客人一定是下一个上桌的客人（不需要绝对公平）。
- 该程序应为想要入座的客人、已经入座的客人和已经离席的客人制作一个屏幕输出。

2.5.2 咒骂与侮辱——Condition ★★

如今，海盗不再用弯刀决斗，而是用咒骂决斗。

任务：
- 启动两个线程，每个线程代表两个海盗，同时命名线程。
- 一个随机的海盗开始咒骂，开始了一场无休止的侮辱比赛。
- 咒骂应从给定的咒骂集合中随机抽取。
- 在咒骂之前，海盗可能需要长达 1 秒的时间"暂停思考"。

2.5.3 从颜料盒中取出笔——Condition ★★

在幼儿园里，小海盗们经常聚在一起画画，可惜只有一个颜料盒中的 12 支笔。当一个孩子从颜料盒里取出笔后，另一个孩子想要取出颜料盒里剩下的笔时就必须等待。

这个场景可以用线程很好地实现。

任务：

- 为一个颜料盒编写类 Paintbox。Paintbox 应该得到一个构造函数，并接受最大数量的空闲的笔。
- Paintbox 类有一个方法 acquisitionPens(int numberOfPens)，孩子们可以用它来请求笔的数量。所需的笔的数量可能大于可用的数量，那么该方法应阻断，直到所需数量的笔再次可用。
- Paintbox 类还有方法 releasePens(int numberOfPens) 用于放回笔。该方法表明笔又可以使用了。
- 创建 Child 类。
- 为 Child 类提供一个构造函数，以便每个孩子都有一个名称并可以获取对颜料盒的引用。
- Child 类应该实现 Runnable 接口。该方法旨在确定一个介于 1 和 10 之间的随机数，它代表所需的笔的数量。然后，孩子询问颜料盒里笔的数量。孩子使用笔 1~3 秒，然后将所有笔放回颜料盒中——不多也不少。之后，孩子等待 1 秒，然后再次开始请求随机数量的笔。

可以让下列孩子开始绘画：

```
public static void main( String[] args ) {
  Paintbox paintbox = new Paintbox( 12 );
  ExecutorService executor = Executors.newCachedThreadPool();
  executor.submit( new Child( "Mirjam", paintbox ) );
  executor.submit( new Child( "Susanne", paintbox ) );
  executor.submit( new Child( "Serena", paintbox ) );
  executor.submit( new Child( "Elm", paintbox ) );
}
```

2.5.4　玩"剪刀石头布"游戏——CyclicBarrier ★★★

"剪刀石头布"是一种古老的游戏，早在 17 世纪人们就开始玩了。发出开始信号后，两名玩家用一只手组成剪刀、石头或布的形状。以下规则决定了哪个玩家获胜：

- 剪刀剪布（剪刀胜）。
- 布包住石头（布胜）。
- 石头使剪刀变钝（石头胜）。

每一个手势都可能输或赢。

我们想为游戏编写一个模拟程序，并使用以下手势列表作为基础：

```java
enum HandSign {
  SCISSORS, ROCK, PAPER;

  static HandSign random() {
    return values()[ ThreadLocalRandom.current().nextInt( 3 ) ];
  }

  int beats( HandSign other ) {
  return (this == other) ? 0 :
        (this == HandSign.ROCK && other == HandSign.SCISSORS
          || this == HandSign.PAPER && other == HandSign.ROCK
          || this == HandSign.SCISSORS && other == HandSign.PAPER) ?
          +1 : -1;
  }
}
```

enum HandSign 声明石头、布、剪刀的三个枚举元素。静态方法 random( ) 返回一个随机手势。beats(HandSign) 方法类似 Comparator 方法：它将当前手势与传递的手势进行比较，如果手势相等，则返回 0；如果自己的手势相比传递的手势为优胜方，则返回 +1，否则返回 –1。

任务：

- 运行一个启动线程，每秒启动"剪刀石头布"游戏。ScheduledExecutorService 可用于重复执行。
- 玩家由一个 Runnable 表示，该 Runnable 选择一个随机手势并将选择放入带有 add(...) 的 ArrayBlockingQueue 类型的数据结构。
- 玩家选择手势后，应在先前创建的 CyclicBarrier 中调用 await( ) 方法。
- CyclicBarrier 的构造函数应该获得一个在游戏结束时确定获胜者的 Runnable。Runnable 使用 poll( ) 从 ArrayBlockingQueue 获取两个手势，比较它们并评估赢家和输家。玩家 1 位于数据结构的第一个位置，玩家 2 位于数据结构的第二个位置。

2.5.5 找到跑得最快的人——CountDownLatch ★★

Bonny Brain 需要跑得快的人参加下一次突袭。因此，她组织了一场比赛，让最好的选手参赛。Bonny Brain 拿着发令枪站在跑道上，所有人都在等待发令信号。

任务：
- 首先，创建 10 个线程，等待来自 Bonny Brain 的信号。然后，线程开始，在自由选择的 10~20 秒之间运行。最后，线程应该把它们的时间写进一个共同的数据结构，这样线程的名称（跑步者的名字）就会与运行时间一起被记录。
- Bonny Brain 在 main 线程中启动跑步者，最后显示所有的跑步时间，并将跑步者的名称升序排列。

> **小提示：**
> 如果要将多个线程集中在一个点上，那么一个好的方法是使用 CountDownLatch。CountDownLatch 用一个整数（一个计数器）初始化，并提供两个中心方法：
> - countDown( ) 使计数器递减。
> - await( ) 用于阻塞，直到计数器变为 0。

2.6 可供参考的解决方案

任务 2.1.1：创建用于招手和挥动旗帜的线程
以下是 main (...) 方法的一部分：

```java
class Wink implements Runnable {
  @Override public void run() {
    for ( int i = 0;i< 50; i++ )
      System.out.printf( "Wink;%s%n", Thread.currentThread() );
  }
}

Runnable winker = new Wink();
Runnable flagWaver = () -> {
  for ( int i = 0;i< 50; i++ )
```

```
    System.out.printf( "Wave flag;%s%n", Thread.currentThread() );
};

Thread winkerThread = new Thread( winker );
Thread flagWaverThread = new Thread( flagWaver, "flag waver" );

winkerThread.start();
flagWaverThread.start();
```
列表 2.1 com/tutego/exercise/thread/CaptainsParade.java

在 run( ) 方法的主体和 Lambda 表达式中，我们发现一个简单的循环与所需的输出。

在创建了 Runnable 实例后，必须将它们连接到线程。为此，Runnables 被传递给 Thread 的构造函数。构造函数多次重载；变体允许设置名称，我们为摇旗者设置名称。构建线程实例并不启动一个线程，为此需要调用 Thread 方法 start( )。

```
Wink; Thread[Thread-0,5,main]
Wave flag; Thread[flag waver,5,main]
Wave flag; Thread[flag waver,5,main]
Wave flag; Thread[flag waver,5,main]
Wink; Thread[Thread-0,5,main]
Wink; Thread[Thread-0,5,main]
Wink; Thread[Thread-0,5,main]
Wink; Thread[Thread-0,5,main]
...
```

输出显示了 Thread 的 toString( ) 表示，它由线程名称、线程优先权和线程组组成。从输出中可以看到，Thread-0（这是自动分配的名称）和 flag waver 交替出现。每次调用时，输出结果都会略有不同——并发程序通常是非确定性的。

任务 2.1.2：不再招手或挥动旗帜：结束线程

```
Runnable winker = () -> {
  while ( ! Thread.currentThread().isInterrupted() ) {
    System.out.printf( "Wink;%s%n", Thread.currentThread() );
    try { TimeUnit.SECONDS.sleep( 2 ); }
    catch ( InterruptedException e ) { Thread.currentThread().interrupt(); }
```

```java
      }
    };

    Runnable flagWaver = () -> {
      while ( ! Thread.currentThread().isInterrupted() ) {
        System.out.printf( "Wave flag;%s%n", Thread.currentThread() );
        try { TimeUnit.SECONDS.sleep( 2 ); }
        catch ( InterruptedException e ) { Thread.currentThread().interrupt(); }
      }
    };

    Thread winkerThread    = new Thread( winker );
    Thread flagWaverThread = new Thread( flagWaver );

    winkerThread.start();
    flagWaverThread.start();

    String message = "Submit 'endw' or 'endf' to end the threads or cancel to end main thread";
    for ( String input;
          ( input = JOptionPane.showInputDialog( message )) != null; ) {
      if ( input.equalsIgnoreCase( "endw" ) )
        winkerThread.interrupt();
      else if ( input.equalsIgnoreCase( "endf" ) )
        flagWaverThread.interrupt();
    }
```

列表 2.2 com/tutego/exercise/thread/CaptainsParadeIsInterrupted.java

为了使线程能够对中断做出反应，必须注意查询 Runnable 中的中断标志。isInterrupted( ) 方法处理这个查询。我们将查询放在一个 while 循环中，因为我们希望只要未设置中断标志，操作就会继续。在两个循环的主体中有一个屏幕输出和 2 秒的等待时间。休眠时有一个特殊之处需要考虑：如果休眠被中断，那么首先，sleep(...) 方法会抛出 InterruptedException；其次，它会重置中断标志。因此，我们必须在 catch 块中再次设置中断标志，并使用 interrupt( ) 方法。这也正是我们在 main(...) 方法中使用的方法，用于通知选定线程自行终止。

使用 stop( ) 粗暴地终止线程和设置标志有一个关键区别：当调用 stop( ) 方法

时，线程是硬终止的，它可以处于任何状态。设置中断标志需要线程的积极参与。线程必须独立请求中断标志并无一例外地终止 run( ) 方法。

任务 2.1.3：参数化 Runnable

run( ) 方法没有返回值，也没有参数列表。因此，run( ) 方法必须以不同的方式获取参数。为此可以使用 run( ) 方法访问变量。

两个建议解决方案如下：

```java
class PrintingRunnable implements Runnable {
  private final String text;
  private final int repetitions;

  PrintingRunnable( String text, int repetitions ) {
    this.text = text;
    this.repetitions = repetitions;
  }

  @Override public void run() {
    for ( int i = 0; i < repetitions; i++ )
      System.out.printf( "%s;%s%n", text, Thread.currentThread() );
  }
}
```

列表 2.3 com/tutego/exercise/thread/ParameterizedRunnable.java

编写一个实现 Runnable 的新类，我们可以给它一个接受不同状态的构造函数。我们可以在对象变量中注意到这些状态。如果我们再调用构造函数，则在创建 Runnable 对象时设置值，如果线程稍后调用 run( ) 方法，则 run( ) 的实现可以访问这些值。

```java
public static Runnable getPrintingRunnable( String text, int repetitions ) {
  return () -> {
    for ( int i = 0; i < repetitions; i++ )
      System.out.printf( "%s;%s%n", text, Thread.currentThread() );
  };
}
```

列表 2.4 com/tutego/exercise/thread/ParameterizedRunnable.java

第二个建议解决方案使用工厂方法。提醒一下，工厂是对象创建者和构造函数的替代品，在这种情况下，参数化不是通过构造函数完成的，而是通过方法完成的。Lambda 表达式可以使用局部变量，参数变量是其中的一部分。一个方法可以在主体中返回一个 Lambda 表达式，因为它可以访问参数，因此我们也用它创建了一个参数化的 Runnable。

任务 2.2.1：使用睡眠线程延迟处理

```java
public class Sleep {

  static long parseSleepArgument( String arg ) {
    Matcher matcher = Pattern.compile( "(\\d+)(\\D)?" ).matcher( arg );
    boolean anyMatch = matcher.find();

    // Check if any match at all or gibberish
    if ( ! anyMatch ) {
      System.err.printf( "sleep: invalid time interval '%s'%n", arg );
      System.exit( 2 );
    }

    // Found at least a number, but maybe too huge to parse
    long seconds = 0;
    try { seconds = Long.parseLong( matcher.group( 1 ) ); }
    catch ( NumberFormatException e ) {
      System.err.printf( "sleep: interval to huge '%s'%n", arg );
      System.exit( 3 );
    }

    // Also a unit?
    String unit = matcher.group( 2 );
    if ( unit == null )
      return seconds;

    switch ( unit ) {
      case "s": break;
      case "m": seconds = TimeUnit.MINUTES.toSeconds( seconds ); break;
```

```java
      case "h": seconds = TimeUnit.HOURS.toSeconds( seconds ); break;
      case "d": seconds = TimeUnit.DAYS.toSeconds( seconds ); break;
      default:
        System.err.printf( "sleep: invalid interval unit '%s'%n", arg );
        System.exit( 4 );
    }

    return seconds;
  }

  public static void main( String[] args ) {
    if ( args.length == 0 ) {
      System.err.println( "sleep: missing operand" );
      System.exit( 1 );
    }

    long seconds = 0;
    for ( String arg : args )
      seconds += parseSleepArgument( arg );
    try { TimeUnit.SECONDS.sleep( seconds ); }
    catch ( InterruptedException e ) { /* intentionally empty */ }
  }
}
```

列表 2.5 com/tutego/exercise/thread/Sleep.java

解析参数占用了大部分空间。我们将解析外包给 parseSleepArgument(String) 方法，该方法以秒为单位返回等待时间。正则表达式有助于识别任意长度的十进制数字，其后还有一个非十进制的字符。这两部分都在圆括号中，这样我们之后就可以通过分组准确地访问该数字和单位。

如果正则表达式不匹配，则会输出错误报告并使用 System.exit(int) 结束程序。如果返回非 0，则表示有错误。无错误的程序通常返回 0。

如果字符串中有两个组，则继续。我们将第一个匹配组转换为秒数。如果数字太大并且不适用于 long 类型，则可能引发异常。通过选择正则表达式，我们排除了数字中的字母，但正则表达式并没有限制大小。NumberFormatException 会在发生错误时提醒我们；同时会有一个输出，程序被终止。

虽然数字被识别，但其后面是否有单位？提取第二个匹配组，如果它为 null，则我们可以退出该方法，因为没有指定单位。如果它不等于 null，那么我们得到一个字符，并检查它是哪个字符。如果是 s，那么中断 switch-case 并且不必进行任何转换。如果它是 m, h 或 d，则 TimeUnit 枚举中的常量有助于将其转换为秒。如果它不是这四个字符之一，则会出现错误报告并终止程序。最好的情况是返回秒数。

main(...) 方法首先执行测试以确定是否传递了任何参数。如果没有，则会出现一条错误报告，程序以 System.exit(...) 和错误代码退出。这比使用 return 退出 main(...) 方法更正确，因为这会导致退出代码为 0，这对 Shell 中的调用程序来说是一个错误信号。

扩展的 for 循环遍历命令行中的所有参数，不需要索引。我们将每个字符串传输到 parseSleepArgument(...) 方法，并将结果相加到变量 seconds 中。最后，sleep(long) 方法使主线程休眠指定的时间（秒）。sleep(...) 方法抛出 InterruptedException，一个我们还需要处理的已检查异常。但是，我们不必在 catch 中写任何东西，因为没有外部线程干扰我们。

任务 2.2.2：通过线程观察文件变化

```java
public class FileChangeWatcher implements Runnable {

  private final Path path;
  private final Consumer<Path> callback;

  public FileChangeWatcher( String filename, Consumer<Path> callback ) {
    this.callback = Objects.requireNonNull( callback );
    path = Paths.get( filename );
  }

  @Override
  public void run() {
    try {
      FileTime oldLastModified = Files.getLastModifiedTime( path );

      while ( true ) {
        TimeUnit.MILLISECONDS.sleep( 500 );
        FileTime lastModified = Files.getLastModifiedTime( path );
        if ( ! oldLastModified.equals( lastModified ) ) {
```

```java
          callback.accept( path );
          oldLastModified = lastModified;
        }
      }
    }
    catch ( Exception e ) {
      // Catch any exception and wrap in a runtime exception
      throw new RuntimeException( e );
    }
  }

  public static void main( String[] args ) {
    Consumer<Path> callback = path -> System.out.println(
      "File changed " + path );
    new Thread( new FileChangeWatcher( "c:/file.txt", callback ) ).start();
  }
}
```

列表 2.6 com/tutego/exercise/thread/FileChangeWatcher.java

由于 Runnable 接口的 run( ) 方法没有参数列表，所以可能的传输或返回值必须以其他方式传输。解决方法很简单：如果我们有一个实现了 Runnable 接口的类，则可以使用一个参数化的构造函数，这样就可以从外面引入一个状态。这正是 FileChangeWatcher 的构造函数所做的——它接受一个文件名和一个消费者。构造函数测试是否错误传输了 null，然后抛出异常，否则构造函数会记住私有对象变量中的状态。Paths 类的工厂方法会自动抛出 NullPointerException，因此我们不必对文件名进行自己的 null 测试。

中央 run( ) 方法由线程调用。首先，我们得到文件最后修改的时间。这是我们稍后将比较的第一个参考点。所有这些都陷入了一个死循环，因为还没有检查或识别和更改它就已经结束了。

循环体以等待开始。由于我们不想用紧密循环引发性能瓶颈，所以我们让线程休眠半秒，然后再次查询上次更改的时间并与旧时间进行比较。如果时间不一样，则文件已更改。在这种情况下，我们必须通知消费者，在传递的回调对象上调用 accept(...) 方法并传递路径。我们应该知道调用是同步和阻塞的。这意味着如果回调方法工作时间长了，我们的线程就不能再监听文件了。

调用消费者后，我们将旧日期设置为当前日期，并在循环中再次开始比较。

有几个地方可能发生异常，这就是为什么在循环和第一次时间查询周围有一个 try-catch 块。首先，输入/输出方法抛出异常，然后休眠。在循环外捕获所有异常，这是一个设计决策。我们也可以采取不同的做法，以便在无限循环中捕获异常，并且即使在发生异常时也会检查文件更改。但是，如果文件被删除，则线程不应该继续。如果有任何异常，那么我们将它们包装在 RuntimeException 中并抛出。这会导致线程终止。被 RuntimeException 中止的线程可以被 Thread.UncaughtExceptionHandler 识别（下一个任务就涉及该内容）。

异常有一个不太显眼的地方是回调操作，它也可以抛出 RuntimeException。不及时处理会导致线程死亡，但我们也不做任何其他事情……我们也可以考虑将自己的异常与 Consumer 中的异常分开，或者使用第二个回调对象报告它们。

任务 2.2.3：捕捉异常

```java
enum GlobalExceptionHandler implements Thread.UncaughtExceptionHandler {
  INSTANCE;

  @Override public void uncaughtException(
      Thread thread, Throwable uncaughtException ) {
    Logger logger = Logger.getLogger( getClass().getSimpleName() );
    logger.log( Level.SEVERE, uncaughtException.getMessage() +
      " from thread " + thread, thread );
  }
}

public class GlobalExceptionHandlerDemo {
  public static void main( String[] args ) throws Exception {
    Thread.setDefaultUncaughtExceptionHandler( GlobalExceptionHandler.INSTANCE );

    Thread zeroDivisor = new Thread( () -> System.out.println( 1 / 0 ) );
    zeroDivisor.start();

    Thread indexOutOfBound =
      new Thread( () -> System.out.println( (new int[0])[1] ) );
    indexOutOfBound.setUncaughtExceptionHandler( ( t, e ) -> {} );
    indexOutOfBound.start();

    new URL( "captain" );
```

```
    }
  }
```

列表 2.7 com/tutego/exercise/thread/GlobalExceptionHandlerDemo.java

输出如下：

```
Juli 04, 2020 2:07:13 PM com.tutego.exercise.thread.GlobalExceptionHandler
uncaughtException
SCHWERWIEGEND: / by zero from thread Thread[Thread-0,5,main]
Juli 04, 2020 2:07:13 PM com.tutego.exercise.thread.GlobalExceptionHandler
uncaughtException
SCHWERWIEGEND: no protocol: captain from thread Thread[main,5,main]
```

enum 实现了 UncaughtExceptionHandler，它与单个静态变量 INSTANCE 一起产生一个单例模式。枚举元素从函数式接口实现 uncaughtException(...) 方法。激活时，JVM 向该方法传递对垂死线程和未处理异常的引用。其类型是 Throwable，这意味着可以报告一个错误。

main(...) 方法为所有线程全局设置 UncaughtExceptionHandler。第一个线程将在除以 0 时抛出 ArithmeticException，我们的全局 UncaughtExceptionHandler 将直接报告。

第二个线程也因异常而中止，但在这里我们通过不报告任何内容的 Lambda 表达式设置了本地 UncaughtExceptionHandler。

主线程也是无意义的，因为 URL 类的构造函数会用这个参数抛出异常。我们没有使用 try-catch 块来拦截和处理它，但是 main(...) 方法将它转发到 JVM。这也激活了设置的 UncaughtExceptionHandler。

任务 2.3.1：使用线程池

以下是 main(...) 方法的一部分，它声明了一个本地 DistributeGift 类并在代码中使用它：

```
class DistributeGift implements Runnable {
  private final String gift;

  public DistributeGift( String gift ) { this.gift = gift; }

  @Override public void run() {
```

```
    try {
      System.out.println( Thread.currentThread().getName() +
        " gives " + gift );
      Thread.sleep( ThreadLocalRandom.current().nextInt( 1000, 4000 ) );
    }
    catch ( InterruptedException e ) { /* Ignore */ }
  }
}

Iterator<String> names = Arrays.asList( "Polly Zist", "Jo Ghurt",
  "Lisa Bonn" ).iterator();

ExecutorService crew = Executors.newCachedThreadPool( runnable -> {
  ThreadFactory threadFactory = Executors.defaultThreadFactory();
  Thread thread = threadFactory.newThread( runnable );
  thread.setName( names.next() );
  return thread;
} );

String[] gifs = { "Dragon", "Pomsies", "Coat", "Tablet", "Doll",
  "Art Station", "Bike", "Card Game", "Slime", "Nerf Blaster" };
for ( String gift : gifs ) {
  Thread.sleep( ThreadLocalRandom.current().nextInt( 1000, 2000 ) );
  crew.submit( new DistributeGift( gift ) );
}
```

列表 2.8 com/tutego/exercise/thread/GiftsInTheOrphanage.java

　　Runnable 是线程执行的操作。在该框架中，Java 为要执行的程序代码提供了两个接口：Runnable 和 Callable。其中，Callable 仅在后台程序想要返回某些内容时使用，这在我们的示例中不是必需的。

　　我们可以使用 Lambda 表达式轻松实现 Runnable，但在这里不行，因为 Runnable 每次都必须单独连接到字符串，即礼物（如果可以通过 Lambda 表达式中的变量获取数据，那么可能有所不同，但 Runnable 本身不应该获取数据）。因此，有一个类实现了 Runnable，并在构造函数中接收要分发的礼物。Runnable 进行屏幕输出，休眠一段时间，然后结束。

下一步，我们创建一个线程池。基本上，Runnable 是由新线程还是现有线程运行都没有关系。与正常的对象创建相比，创建线程的成本要高得多。无参数方法 Executors.newCachedThreadPool( ) 可用于创建线程池，或者我们可以自己创建和参数化线程池的一个特殊变体。本任务并没有要求这样做，这是出于外观上的考虑，因为这样可以设置线程的名称。为了能够尽可能多地接管基础设施，我们查询 defaultThreadFactory( )，通过 newThread(...) 创建一个线程，然后可以使用已知的 Thread 方法 setName(...) 设置线程的名称。船员的姓名是通过 next( ) 方法从 Iterator 中获取的，由于该任务需要线程池中的线程不超过 3 个，所以不会抛出异常。

最后一部分是工作包的启动。程序遍历数组，总是用给定的礼物创建新的 Runnable 对象。submit(...) 将 Runnable 传递给线程池，线程池选择一个空闲的线程或在开始时创建一个，从而分发礼物。

任务 2.3.2：确定网站的最后修改

```
class WebResourceLastModifiedCallable implements Callable<ZonedDateTime> {

  private final URL url;

  WebResourceLastModifiedCallable( URL url ) {
    this.url = url;
  }

  @Override public ZonedDateTime call() throws IOException {
    HttpURLConnection con = (HttpURLConnection) url.openConnection();
    long dateTime = con.getLastModified();
    con.disconnect();
    return ZonedDateTime.ofInstant( Instant.ofEpochMilli( dateTime ),
      ZoneId.of( "UTC" ) );
  }
}
```

列表 2.9 com/tutego/exercise/thread/PageLastModifiedCallableDemo.java

Callable 接口的实现并不令人意外。参数化构造函数接收一个 URL 对象并将其存储在一个私有变量中，以便以后可以在 call( ) 方法中访问该 URL。call(...) 方法的实现与模板没有太大区别，只是我们可以轻松地从 call( ) 方法向上转发异常，而不是捕获异常。如果随后发生异常，则 get(...) 会被 ExecutionException 中断。

用法如下所示：

```java
ExecutorService executor = Executors.newCachedThreadPool();
URL url = new URL( "https://en.wikipedia.org/wiki/Main_Page" );
Callable<ZonedDateTime> callable = new WebResourceLastModifiedCallable( url );

Future<ZonedDateTime> dateTimeFuture = executor.submit( callable );

try {
  System.out.println( executor.submit( callable ).get( 1,
    TimeUnit.MICROSECONDS ) );
}
catch ( InterruptedException | ExecutionException | TimeoutException e ) {
  e.printStackTrace();
}

try {
  ZonedDateTime wikiChangedDateTime = dateTimeFuture.get();
  System.out.println( wikiChangedDateTime );
  System.out.println( Duration.between( wikiChangedDateTime,
    ZonedDateTime.now( ZoneId.of( "UTC" ) ) ).toMinutes() );
}
catch ( InterruptedException | ExecutionException e ) {
  e.printStackTrace();
}

executor.shutdown();
```

列表 2.10 com/tutego/exercise/thread/PageLastModifiedCallableDemo.java

为了发送 Callable 实例，我们使用通过 newCachedThreadPool( ) 获得的线程池。URL 对象指定维基百科主页，用这个 URL 创建 WebResourceLastModifiedCallable 后，我们可以将 Callable 对象放在 submit(..) 方法中。仅此一项不会触发任何异常（除了参数化的 URL 构造函数），仅当使用 get(...) 检索结果时才需要异常处理。

该示例实现了两个场景：第一个 submit(...) 从线程池执行一个 Callable，然后它在后台工作；第二个 submit(...) 也将一个 Callable 放入线程池中，但只给 get(...) 1 微秒的时间来完成，当然这还不够，还会抛出 TimeoutException。

第二个 try-catch 块用于第一个发送的 Callable。使用 get( ) 以阻塞方式向 dateTimeFuture 询问结果，运行时环境可能已经用了几毫秒来处理先前的查询。我们比较一下 catch 块就会发现，如果有时间限制，则必须处理一个额外的 TimeoutException。然而，无参数的 get( ) 的情况并非如此；异常仍然为 InterruptedException 和 ExecutionException。

为了计算差值，我们使用类 Duration 的 between(...) 方法。该静态方法接收两个参数：来自 Future 的时间和当前的时间。别忘记，不要只使用 now( ) 获取当前时间，还必须获取 UTC 时区的当前时间，否则差是错误的，除非应用程序本身位于 UTC ± 0 时区。

任务 2.4.1：在诗集里写下回忆

问题是两个线程访问共享资源。两者都同时访问资源并弄乱了输出。输出可能如下所示：

```
DWeinen Bdulu mlaenc hbsrt,au lcahcenhe Sno nsneien saclhelei.n Wuennd n
icduh w ebriansuct,h wCeainpastta diun aClileoiCinaeo
_,.-'~'-.,__,.-'~'-.,__,.-'~'-.,__,.-'~'-.,__,.-'~'-.,_
zum Fröhlichsein
_,.-'~'-.,__,.-'~'-.,__,.-'~'-.,__,.-'~'-.,__,.-'~'-.,_
```

添加字母的原因是需要保护的临界区。synchronized 块已经过时，我们想使用 Lock 对象。ReentrantLock 实现适用于我们的例子。

```
Lock lock = new ReentrantLock( true );

class Author implements Runnable {
  private final String text;
  private final FriendshipBook book;

  public Author( String text, FriendshipBook book ) {
    this.text = text;
    this.book = book;
  }

  @Override public void run() {
    try {
```

```java
        lock.lock();
        for ( int i = 0; i < text.length(); i++ ) {
          book.appendChar( text.charAt( i ) );
          Thread.sleep( 1 );
        }
        book.appendDivider();
      }
      catch ( InterruptedException e ) { /* Ignore */ }
      finally {
        lock.unlock();
      }
    }
  }
}
```

列表 2.11 com/tutego/exercise/thread/WriteInFriendshipBook.java

　　线程通过 Lock 对象进行协调。当一个线程进入临界区时，另一个线程将不得不等待，直到临界区被释放。在 Lock 中，有两个主要方法用于进入和离开块：lock( ) 和 unlock( )。ReentrantLock 的构造函数被重载，参数化的变体通过一个 boolean 参数决定 Lock 是否公平。在这种情况下，公平意味着等待时间最长的线程被允许首先进入被释放的块，否则行为是不确定的。这也是它与关键词 synchronized 的区别之一，它允许 JVM 选择任何线程。分配是否公平取决于 JVM 的实现，synchronized 的公平性是不可控的。

　　建议解决方案稍微调整了源代码。一处在类声明 Autor 之外，因为 Lock 对象必须对所有线程都可用，线程使用它进行协调。另一处在 run( ) 方法内部，因为保存的操作是写入诗集。在循环之前，使用 lock( ) 关闭块，然后附加所有字符，写入分隔符，最后再次调用 unlock( )，为下一个线程释放块。unlock( ) 应该始终位于 finally 块中，因为如果请求 Lock，则应该始终释放它，即使存在 return 或退出该方法的异常。finally 块总是被处理，不管是否存在异常。

　　除了 lock( ) 方法之外，还有第二种方法 lockInterruptibly( )，它可以被外部的中断打断。lock( ) 不响应外部的中断，这意味着 InterruptedException 的 catch 仅适用于 sleep(...) 方法。

任务 2.5.1：与船长一起参加宴会——Semaphore

Folgendes ist Teil der main(…)-Methode:

```
Semaphore seats = new Semaphore( 6 - 2 );
```

```java
class Guest implements Runnable {
  private final String name;
  public Guest( String name ) { this.name = name; }

  @Override
  public void run() {
    try {
      System.out.printf( "%s is waiting for a free place%n", name );
      seats.acquire();
      System.out.printf( "%s has a seat at the table%n", name );
      Thread.sleep( ThreadLocalRandom.current().nextInt( 2000, 5000 ) );
    }
    catch ( InterruptedException e ) { /* Ignore */ }
    finally {
      System.out.printf( "%s leaves the table%n", name );
      seats.release();
    }
  }
}

List<String> names = new ArrayList<>( Arrays.asList(
    "Balronoe", "Xidrora", "Zobetera", "Kuecarro", "Bendover", "Bane",
    "Cody", "Djarin", "Enfy" ) );
for ( int i = 0, len = names.size(); i < len; i++ )
  Collections.addAll( names, "Admiral " + names.get( i ), "Commander " +
    names.get( i ) );

ExecutorService executors = Executors.newCachedThreadPool();

for ( String name : names )
  executors.execute( new Guest( name ) );

executors.shutdown();
}
```

列表 2.12 com/tutego/exercise/thread/Banquette.java

Banquet 类的核心由四部分组成：

1. 创建 Semaphore；
2. 声明客人（一个使用 Semaphore 的 Runnable）；
3. 创建名称；
4. 创建 Runnable 实例并通过线程池执行。

由于餐桌上有 6 个座位，而 Bonny Brain 和 CiaoCiao 船长已经占据了 2 个，所以还剩下 4 个，不同的客人可以互相切换。

所有客人都有一个在对象中注明的姓名。姓名在输出中出现了 3 次。

1. 客人的姓名在开始给出，因为客人正在等待，可能还没有座位。线程必须首先使用 acquire( ) 从 Semaphore 获得许可，并且当已经达到 4 个空闲座位的最大数量时阻塞该方法。屏幕显示一开始每位客人都在等待空位，前 4 位客人立即入座。
2. 解锁 acquire( ) 后，该线程可以和其他 4 个线程一起在桌子上与船长们共度时光。
3. 等待一段时间后，finally 块被处理，并再次输出客人的姓名，因为客人正在离开餐桌。使用 Semaphore，调用 release( ) 方法很重要，这样其他等待的客人才可以就座。

一个简单的算法生成演示数据。一个列表预先分配了一些字符串。此列表现在以原始长度运行，并生成以 Admiral 开头的新字符串和以 Commander 开头的其他字符串。

现在只需要创建线程并且使用 ExecutorService 就可以启动它们。

任务 2.5.2：咒骂与侮辱——Condition
该代码是 main(...) 方法的一部分：

```
Lock lock = new ReentrantLock();
Condition condition = lock.newCondition();

class Insulter implements Runnable {
  private final String[] insults;
  public Insulter( String[] insults ) {
    this.insults = insults;
```

```java
        }

        @Override public void run() {
            while ( Thread.currentThread().isInterrupted() ) {
                try {
                    lock.lock();
                    Thread.sleep( ThreadLocalRandom.current().nextInt( 1000 ) );
                    String name = Thread.currentThread().getName();
                    int rndInsult = ThreadLocalRandom.current().nextInt( insults.length );
                    System.out.println( name + ": " + insults[ rndInsult ] + '!' );
                    condition.signal();
                    condition.await();
                }
                catch ( InterruptedException e ) { Thread.currentThread().interrupt(); }
                finally {
                    lock.unlock();
                }
            }
        }
    }

    String[] insults1 = {
        "Trollop", "You have the manners of a trump",
        "You fight like a cow cocky", "Prat",
        "Your face makes onions cry",
        "You are so full of s**t, the toilet's jealous"
    };
    String[] insults2 = {
        "Wazzock", "I've spoken with rats more polite than you",
        "Chuffer", "You make me want to spew",
        "Check your lipstick before you come for me",
        "You are more disappointing than an unsalted pretzel"
    };
```

```
new Thread( new Insulter( insults1 ), "pirate-1" ).start();
new Thread( new Insulter( insults2 ), "pirate-2" ).start();
```
列表 2.13 com/tutego/exercise/thread/InsultSwordFighting.java

该解决方案使用 Condition，它有两个重要的方法：通知其他等待线程和等待通知。

在我们的例子中，Insulter 是一个 Runnable，它通过具有不同咒骂的构造函数进行初始化。run( ) 方法包含一个无限循环，原则上可以通过中断来终止，因为侮辱永远不会结束，每个动作都会跟着一个反应。Condition 对象只能在 lock( ) 标记临界区时使用——Condition 对象来自 Lock 对象。

经过思考后，选择并输出一个随机咒骂，然后通过 signal( ) 通知另一个线程。使用接下来的 await( ) 再次在线程中等待来自另一个线程的信号。

流程如下。main(...) 启动两个 Insulter 线程，两个线程中的一个将通过 Lock 对象率先进入该区域，然后锁定并声明该区域为自己所有。第二个线程稍后出现，挂在 Lock 中并且不进一步操作。然后，第一个线程开始咒骂，并向另一个等待线程发出信号，表示已结束。发出信号后，第一个线程也等待信号。重要的是要理解，关联的锁是暂时释放的，是为了让其他线程进入受保护的块，接收信号，然后执行动作。当一个线程在 await( ) 中等待时，另一个线程执行它的操作，然后再次设置一个信号。这唤醒了等待线程，该线程重新获得锁并在另一方等待时继续其操作。

任务 2.5.3：从颜料盒中取出笔——Condition

我们从 Paintbox 类开始。用笔的最大数量初始化对象变量 freeNumberOfPens，随后增加或减少。

```
class Paintbox {

  private int freeNumberOfPens;
  private final Lock lock = new ReentrantLock();
  private final Condition condition = lock.newCondition();

  public Paintbox( int maximumNumberOfPens ) {
    freeNumberOfPens = maximumNumberOfPens;
    System.out.printf( "Paintbox equipped with%s pens%n", freeNumberOfPens );
  }
```

```java
    public void acquirePens( int numberOfPens ) {
      try {
        lock.lock();

        while ( freeNumberOfPens < numberOfPens ) {
          System.out.printf( "%d pens from paintbox requested, available only%d, someone has to wait :(%n",
                             numberOfPens, freeNumberOfPens );
          condition.await();
        }

        freeNumberOfPens -= numberOfPens;
      }
      catch ( InterruptedException e ) { Thread.currentThread().interrupt(); }
      finally {
        lock.unlock();
      }
    }

    public void releasePens( int numberOfPens ) {
      try {
        lock.lock();
        freeNumberOfPens += numberOfPens;
        condition.signalAll();
      }
      finally {
        lock.unlock();
      }
    }
}
```

列表 2.14 com/tutego/exercise/thread/Kindergarten.java

acquirePens(int numberOfPens) 由想要从颜料盒中取出一定数量的笔的孩子调用。该操作发生在临界区，因此 Lock 为其他线程锁定该部分。循环条件检查空置笔的数量是否小于所需的数量，如果是，则必须等待。如果信号稍后出现，则必须反复询问此条件是否仍然适用。使用条件判断而不是循环属于编程错误。在

while 循环结束后，空置笔的数量减少，临界区再次被释放。注意：等待信号时，锁会暂时解除。

releasePens(int numberOfPens) 更简单：它增加空置笔的数量，然后向其他等待线程发出信号，说明笔再次可用。signalAll( ) 向所有等待的线程发出信号。就像等待一样，信号必须发生在一个锁定的部分，它通常以 unlock( ) 结束 finally 块。

Child 类实现 Runnable 并在构造函数中获取孩子的姓名和颜料盒。

```java
class Child implements Runnable {
  private final String name;
  private final Paintbox paintbox;

  public Child( String name, Paintbox paintbox ) {
    this.name = name;
    this.paintbox = paintbox;
  }

  @Override
  public void run() {
    while ( ! Thread.currentThread().isInterrupted() ) {
      int requiredPens = ThreadLocalRandom.current().nextInt( 1, 10 + 1 );
      paintbox.acquirePens( requiredPens );
      System.out.printf( "%s got%d pens%n", name, requiredPens );

      try {
        TimeUnit.MILLISECONDS.sleep(
          ThreadLocalRandom.current().nextInt( 1000, 3000 ) );
      }
      catch ( InterruptedException e ) { Thread.currentThread().interrupt(); }

      paintbox.releasePens( requiredPens );
      System.out.printf( "%s returned%d pens%n", name, requiredPens );

      try {
        TimeUnit.SECONDS.sleep(
          ThreadLocalRandom.current().nextInt( 1, 5 + 1 ) );
```

```
          }
          catch ( InterruptedException e ) { Thread.currentThread().
interrupt(); }
        }
      }
    }
```

列表 2.15 com/tutego/exercise/thread/Kindergarten.java

原则上，run( )方法无止境地运行，除非线程被外部中断终止。以下是while循环体的内容：

1. 为所需的笔数指定一个随机数。
2. 在颜料盒中调用 acquirePens(...) 来获取笔的数量。如果所需的数量不可用，则此处可能出现等待情况。
3. 如果笔可用，则 acquirePens(...) 返回并输出获取的笔数。
4. 等待，因为孩子在绘画。
5. 笔被 releasePens(...) 归还。还笔不会阻塞很长时间。
6. 再次等待，然后可以开始新的循环运行。

任务 2.5.4：玩"剪刀石头布"游戏——CyclicBarrier

该代码是 main(...) 方法的一部分：

```
Queue<HandSign> handSigns = new ArrayBlockingQueue<>( 2 );

Runnable determineWinner = () -> {
  HandSign handSign1 = handSigns.poll();
  HandSign handSign2 = handSigns.poll();

  switch ( handSign1.beats( handSign2 ) ) {
    case 0:
      System.out.printf( "Tie, both players choose%s%n", handSign1 );
      break;
    case +1:
      System.out.printf( "Player 1 wins with%s, player 2 loses with%s%n",
                         handSign1, handSign2 );
      break;
```

```
      case -1:
        System.out.printf( "Player 2 wins with%s, player 1 loses with%s%n",
                           handSign2, handSign1 );
        break;
    }
  };

  CyclicBarrier barrier = new CyclicBarrier( 2, determineWinner );

  Runnable playScissorsRockPaper = () -> {
    try {
      handSigns.add( HandSign.random() );
      barrier.await();
    }
     catch ( InterruptedException | BrokenBarrierException e ) { /*
Ignore */ }
  };

  ScheduledExecutorService executor = Executors.newScheduledThreadPool( 2 );
  executor.scheduleAtFixedRate( () -> {
    System.out.println( "Schnick, Schnack, Schnuck" );
    executor.execute( playScissorsRockPaper );
    executor.execute( playScissorsRockPaper );
  }, 0, 1, TimeUnit.SECONDS );
```

列表 2.16 com/tutego/exercise/thread/RockPaperScissorsHandGame.java

在建议解决方案中，我们有两种不同的 Runnable 类型。一个是 Runnable，用于宣布获胜者；另一个是 Runnable，用于执行手势。Runnable 使用 Lambda 表达式实现并访问通用的 handSigns 数据结构。每个玩家随机做一个手势，放进 Queue，等待屏障结束。

屏障本身使用尺寸 2 和 Runnable determineWinner 进行初始化，当在屏障上两次调用 await( ) 时，它总是会执行。determineWinner 将两个手势从队列中取出，使用 beats(...) 来确定哪个玩家得分，并输出有关赢家、输家或平局的控制台消息。

ScheduledExecutorService 帮助每秒重复游戏，并执行 playScissorsRockPaper 后面的 Runnable 两次。

任务 2.5.5：找到跑得最快的人——CountDownLatch

Folgender Code ist in der main(…)-Methode:

```
final int NUMBER_OF_ATHLETES = 10;
CountDownLatch startLatch = new CountDownLatch( 1 );
CountDownLatch endLatch   = new CountDownLatch( NUMBER_OF_ATHLETES );

ConcurrentNavigableMap<Integer, String> records =
  new ConcurrentSkipListMap<>();

Runnable athlete = () -> {
  try {
    startLatch.await();
    int time = ThreadLocalRandom.current().nextInt( 1_000, 2_000 );
    TimeUnit.MILLISECONDS.sleep( time );
    records.put( time, Thread.currentThread().getName() );
    endLatch.countDown();
  }
  catch ( InterruptedException e ) { /* Ignore */ }
};

for ( int i = 0; i < NUMBER_OF_ATHLETES; i++ )
  new Thread( athlete, "athlete-" + (i + 1) ).start();

// Start the race
startLatch.countDown();

// Wait for race to end
endLatch.await();

records.forEach( ( time, name ) -> System.out.printf( "%s in%d ms%n",
  name, time ) );
```

列表 2.17 com/tutego/exercise/thread/SprintRace.java

该建议解决方案使用两个 CountDownLatch 对象。我们将第一个 startLatch 初始化为 1，当多个线程启动时，它们等待这个开始的 CountDownLatch 变为 0。我们将第二个 CountDownLatch endLatch 初始化为 10，即运动员的数量，每当运动员到达终点线时，endLatch 就会减少。

我们使用排序的关联映射按运行时排序。由于线程同时写入数据结构，所以数据结构必须支持这种类型的访问。java.util.concurrent 包为此提供了各种数据结构。我们使用 ConcurrentSkipListMap，这是一种可以处理并发访问的相连存储器。该类实现了 ConcurrentNavigableMap 接口。

运动员是通过 Thread 类的构造函数获得简单名称的线程。线程启动后，它们都在 startLatch 等待到达 0。调用 startLatch.countDown( ) 允许所有运动员线程开始运行。每个线程等待随机的时间，然后将自己写入数据结构并减少 endLatch 中的计数器。主程序还使用 endLatch.await( ) 等待解除封锁，这是通过所有运动员递减计数器使其变为 0 来实现的。运行结束后，forEach(...) 遍历相连存储器并返回时间和姓名。

第 3 章
数据结构和算法

数据结构在应用程序中存储重要信息,通过列表、集合、队列和关联映射来组织。在 Java 中,有关数据结构的接口和类被称为集合 API。由于有如此多的类型可供选择,所以本章的目的是让混乱的局面变得有序,并通过输出来阐明相应集合的用途。

本章使用的数据类型如下:

- java.util.Collection (https://docs.oracle.com/en/java/javase/11/docs/api/java.base/java/util/Collection.html)
- java.util.Collections (https://docs.oracle.com/en/java/javase/11/docs/api/java.base/java/util/Collections.html)
- java.util.Iterable (https://docs.oracle.com/en/java/javase/11/docs/api/java.base/java/lang/Iterable.html)
- java.util.List (https://docs.oracle.com/en/java/javase/11/docs/api/java.base/java/util/List.html)
- java.util.ArrayList (https://docs.oracle.com/en/java/javase/11/docs/api/java.base/java/util/ArrayList.html)
- java.util.LinkedList (https://docs.oracle.com/en/java/javase/11/docs/api/java.base/java/util/LinkedList.html)
- java.util.ListIterator (https://docs.oracle.com/en/java/javase/11/docs/api/java.base/java/util/ListIterator.html)
- java.util.Set (https://docs.oracle.com/en/java/javase/11/docs/api/java.base/java/util/Set.html)
- java.util.HasSet (https://docs.oracle.com/en/java/javase/11/docs/api/java.base/java/util/HashSet.html)

- java.util.TreeSet (https://docs.oracle.com/en/java/javase/11/docs/api/java.base/java/util/TreeSet.html)
- java.util.LinkedHashSet (https://docs.oracle.com/en/java/javase/11/docs/api/java.base/java/util/LinkedHashSet.html)
- java.util.SortedSet (https://docs.oracle.com/en/java/javase/11/docs/api/java.base/java/util/SortedSet.html)
- java.util.Map (https://docs.oracle.com/en/java/javase/11/docs/api/java.base/java/util/Map.html)
- java.util.HashMap (https://docs.oracle.com/en/java/javase/11/docs/api/java.base/java/util/HashMap.html)
- java.util.TreeMap (https://docs.oracle.com/en/java/javase/11/docs/api/java.base/java/util/TreeMap.html)
- java.util.SortedMap (https://docs.oracle.com/en/java/javase/11/docs/api/java.base/java/util/SortedMap.html)
- java.util.WeakHashMap (https://docs.oracle.com/en/java/javase/11/docs/api/java.base/java/util/WeakHashMap.html)
- java.util.Properties (https://docs.oracle.com/en/java/javase/11/docs/api/java.base/java/util/Properties.html)
- java.util.Queue (https://docs.oracle.com/en/java/javase/11/docs/api/java.base/java/util/Queue.html)
- java.util.Deque (https://docs.oracle.com/en/java/javase/11/docs/api/java.base/java/util/Deque.html)
- java.util.BitSet (https://docs.oracle.com/en/java/javase/11/docs/api/java.base/java/util/BitSet.html)
- java.util.concurrent.SynchronousQueue (https://docs.oracle.com/en/java/javase/11/docs/api/java.base/java/util/concurrent/SynchronousQueue.html)

3.1 集合 API 的接口

这次使用的类列表很长。但是，设计遵循一个基本原则，所以它并不复杂。

- 接口描述了数据结构的功能，即"提供什么"。
- 类使用不同的策略来实现来自接口的指令；它们代表"如何实施"。
- 作为开发人员，我们需要了解接口和实现，让我们再复习一下，看看本章中要经常遇到的重点类型（见图 3.1）。

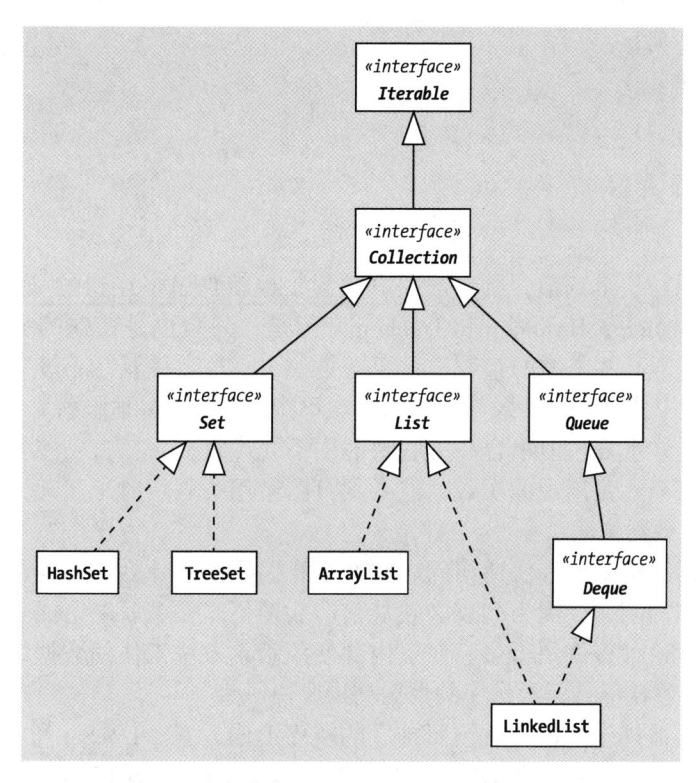

图 3.1 所选数据结构和类型关系的 UML 图示

需要注意的是：

- Iterable 是最通用的接口，表示可以运行什么；Iterable 返回 Iterator 实例。Iterable 不仅是数据结构。
- Collection 是真正代表数据结构的顶级接口。它规定了向集合添加或删除元素的方法。
- 原本的抽象都在 Collection 下，无论是列表、集合还是队列。实现也在其下。
- 一些操作不在数据类型本身上，而是外包给 Collections 类。这同样适用于数组，其中还有一个实用程序类 Arrays。

我们想为类和接口 java.util.Set，java.util.List，java.util.Map，java.util.HashSet，java.util.TreeSet，java.util.Hashtable，java.util.HashMap 和 java.util.TreeMap 创建决策树。在进行选择时必须考虑以下因素：

- 密钥访问；
- 允许重复；
- 快速访问；
- 排序迭代；
- 线程安全。

如果访问是从键到值，则一般关联映射，也就是 Map 接口的一种实现。Map 的实现有 Hashtable，HashMap 和 TreeMap。但是，列表也是特殊的关联映射，其中索引是从 0 开始并递增的整数。只要键是一个小整数并且几乎没有空白，列表就可以很好地工作。任意整数与对象的关联不能用列表很好地映射。

列表中允许重复，但集合和关联映射中不允许重复。存在一些要求，一个集合应该注明一个元素出现的次数，但必须自己使用关联映射来实现，该关联映射将该元素与计数器关联。

所有数据结构都允许快速访问。问题是我们要询问什么。列表不能快速回答元素是否存在的问题，因为列表必须从前向后遍历。在关联映射或集合中，数据的内部组织使这种查询快得多。对于内部使用哈希方法工作的数据结构，这种存在性测试的回答比保持元素排序的数据结构要快一些。

列表可以被排序，运行时按照排序的顺序返回元素。TreeSet 和 TreeMap 也会根据一个标准进行排序。具有哈希方法的数据结构不按用户的定义排序。

数据结构分为三组：自 Java 1.0 起的数据结构、自 Java 1.2 起的数据结构和自 Java 5 起的数据结构。数据结构 Vector，Hashtable，Dictionary 和 Stack 是在第一个 Java 版本中引入的。这些数据结构都是线程安全的，但它们今天已不再使用。集合 API 是在 Java 1.2 中引入的，所有数据结构都不是线程安全的。新包 java.util.concurrent 是在 Java 5 引入的，那里的所有数据结构都可以防止并发更改。

3.1.1 测试：搜索 StringBuilder ★★

如果以下程序在一个类的 main(...) 方法中，那么它的输出是什么？

```
Collection<String> islands1 = new ArrayList<>();
islands1.add( "Galápagos" );
islands1.add( "Revillagigedo" );
islands1.add( "Clipperton" );
System.out.println( islands1.contains( "Clipperton" ) );
```

```
Collection<StringBuilder> islands2 = new ArrayList<>();
islands2.add( new StringBuilder( "Galápagos" ) );
islands2.add( new StringBuilder( "Revillagigedo" ) );
islands2.add( new StringBuilder( "Clipperton" ) );
System.out.println( islands2.contains( new StringBuilder( "Clipperton" ) ) );
```

如何解释输出？它是否可能偏离了假设的行为？会不会是由于 String 满足了一个重要的要求，而 StringBuilder 没有？

3.2 列表

在任务中，我们想从最简单的数据结构——列表开始。列表是信息的序列，当新的元素被添加时，其顺序被保持，元素可以多次出现。它甚至允许 null 作为一个元素。

3.2.1 唱歌做饭：运行列表，检查特性 ★

CiaoCiao 船长正在组建一个新的团队。团队中的每个人都有一个姓名和一个职业：

```
class CrewMember {
  enum Profession { CAPTAIN, NAVIGATOR, CARPENTER, COOK, MUSICIAN, DOCTOR }
  String name;
  Profession profession;
  CrewMember( String name, Profession profession ) {
    this.name = name;
    this.profession = profession;
  }
}
```

CiaoCiao 船长希望船员中厨师和乐手一样多。

任务：

编写方法 areSameNumberOfCooksAndMusicians(List<CrewMember>)，在厨师和乐手一样多的时候返回 true，否则返回 false。

举例:

```
CrewMember captain   = new CrewMember( "CiaoCiao",
  CrewMember.Profession.CAPTAIN );
CrewMember cook1     = new CrewMember( "Remy", CrewMember.Profession.COOK );
CrewMember cook2     = new CrewMember( "The Witch Cook",
  CrewMember.Profession.COOK );
CrewMember musician1 = new CrewMember( "Mahna Mahna",
  CrewMember.Profession.MUSICIAN );
CrewMember musician2 = new CrewMember( "Rowlf",
  CrewMember.Profession.MUSICIAN );

List<CrewMember> crew1 = Arrays.asList( cook1, musician1 );
System.out.println( areSameNumberOfCooksAndMusicians( crew1 ) ); // true

List<CrewMember> crew2 = Arrays.asList( cook1, musician1, musician2,
  captain );
System.out.println( areSameNumberOfCooksAndMusicians( crew2 ) ); // false

List<CrewMember> crew3 = Arrays.asList( cook1, musician1, musician2,
  captain, cook2 );
System.out.println( areSameNumberOfCooksAndMusicians( crew3 ) ); // true
```

3.2.2 从列表中过滤评论★

Bonny Brain 正在阅读迪普图鲁斯·迪姆维特船长的一本旧日志，其中反复提到四个条目：

1. 罗盘航向；
2. 水流速度；
3. 天气；
4. 评论和一般观察。

Bonny Brain 正在寻找评论中的一个特定条目，因此要从一个字符串列表中删除第一、第二和第三个条目，以便只保留有评论的第四个条目。

任务：
实现方法 void reduceToComments(List<String> lines) 来删除传递列表中的第一、第二和第三个条目，只保留第四个条目。

举例：
- "A1" "A2" "A3" "A4" "B1" "B2" "B3" "B4" "C1" "C2" "C3" "C4" → "A4" "B4" "C4"；
- 空列表 → 无事发生；
- "A1" → 异常 Illegal size 1 of list, must be divisible by 4。

3.2.3 缩短列表，因为衰退不存在★

对于 CiaoCiao 船长来说，一切只能上升；当他阅读数字序列时，它们也只能增加。

任务：
- 编写方法 trimNonGrowingNumbers(List<Double> numbers) 来截断列表。如果下一个数字不再大于或等于前一个数字，则截断该列表。
- 思考：传递的列表必须是可修改的，以便可以删除元素。

举例：
- 如果列表中包含数字 1，2，3，4，5，那么列表保持原样。
- 如果列表中包含数字 1，2，3，2，1，那么列表就被缩短为 1，2，3。

3.2.4　和朋友一起吃饭：比较元素，找到共同点★

Bonny Brain 正计划在陆地举办一场派对，所有客人围在一起坐成大圈，相邻的客人应该至少有一个共同点。客人按以下类型声明：

```
public class Guest {
  public boolean likesToShoot;
  public boolean likesToGamble;
  public boolean likesBlackmail;
}
```

任务：
- 编写方法 int allGuestsHaveSimilarInterests(List<Guest> guest)，如果所有客人与邻居至少有一个相同的属性，则返回 −1。否则，返回 >= 0 并且索引正好在第一个坐错的客人上，也就是说，他与任何邻居都没有共同点。
- Guest 类可以任意扩展。

3.2.5　检查列表的相同元素顺序★

胖子唐尼骨刺和盟友里昂潜入匿名海盗支持小组，他们应该向 CiaoCiao 船长报告，在谈话圈中谁坐在谁的右边。两人都试图记住这些。他们在枚举的时候不一定从同一个人开始。

任务：
- 编写方法 isSameCircle(List<String> names1, List<String> names2)，该方法测试两个列表中的姓名是否以相同的顺序紧跟在一起。记住，人们坐成一圈，名单上的最后一个人"坐在"名单上第一个人的旁边。

举例：
- 列表 1：Alexandre, Charles, Anne, Henry；列表 2：Alexandre, Charles, Anne, Henry → 一致。
- 列表 1：Anne, Henry, Alexandre, Charles；列表 2：Alexandre, Charles, Anne, Henry → 一致。
- 列表 1：Alexandre, Charles, Anne, Henry；列表 2：Alexandre, Charles, Henry, Anne → 不一致。
- 列表 1：Anne, Henry, Alexandre, Charles；列表 2：Alexandre, William, Anne, Henry → 不一致。

3.2.6 现在播报天气：寻找重复元素 ★

拿破仑·纳斯正在和 Bonny Brain 谈论天气："过去几个月连续下了这么多天雨，这对捕鱼者来说很不利。"Bonny Brain 回答："这些下雨天并不是连续的！"谁是正确的？

给出一个天气数据列表：

Regen, Sonne, Regen, Regen, Hagel, Schnee, Sturm, Sonne, Sonne, Sonne, Regen, Regen, Sonne

Sonne（太阳）在列表中连续出现了 3 次。这是我们想知道的。虽然 Regen（雨）在列表中出现的频率更高，但这与解决方案无关。

任务：
- 创建新类 WeatherOccurrence，它是一个存放天气信息的小容器。它是这样实施的：
  ```
  class WeatherOccurrence {
    String  weather;
    int     occurrences;
    int     startIndex;
  }
  ```
- 实现一个签名为 WeatherOccurrence longestSequenceOfSameWeather(List<String> weather) 的方法，它告诉我们：
  - 现在是什么天气；
  - 天气在列表中连续出现的频率；
  - 最长的列表从哪开始。

如果一个天气连续出现的频率一样，则方法可以自由决定返回什么。元素可以为 null。

3.2.7 创建收据输出 ★

一张收据包含条目和数量、产品名称、单价、总价等信息。在这项任务中编辑一个收据。

任务：
- 为收据创建一个新类 Receipt。
- 一张收据由类型 Item 的条目组成。把类 Item 作为嵌套类型放在 Receipt 中。
- 每个 Item 都有一个名称和一个以分为单位的（毛）价格。
- Receipt 应该重写 toString( ) 并返回一个格式化的收据：
  - 输出所有产品和总数。
  - 条目可以在收据上出现多次，应合并计算。例如不是 Nüsse, Nüsse 连在一起，而是 2x Nüsse。条目必须具有相同的名称和价格才相等（Nüsse：坚果）。
  - 使用 NumberFormat.getCurrencyInstance(Locale.GERMANY) 来格式化货币……

举例：
- 创建：

```
Receipt receipt = new Receipt();
receipt.addItem( new Receipt.Item( "Peanuts", 222 ) );
receipt.addItem( new Receipt.Item( "Lightsaber", 19999 ) );
receipt.addItem( new Receipt.Item( "Peanuts", 222 ) );
receipt.addItem( new Receipt.Item( "Log book", 1000 ) );
receipt.addItem( new Receipt.Item( "Peanuts", 222 ) );
System.out.println( receipt );
```

输出如下：

```
3x   Peanuts          2,22 €     6,66 €
1x   Lightsaber     199,99 €   199,99 €
1x   Log book        10,00 €    10,00 €

Sum: 216,65 €
```

3.2.8 测试：装饰数组★

数组在 API 中很常见，并且转换为更方便的 List 也很常见。如果我们在 Java 程序中编写以下内容，那么结果是什么？

```
Arrays.asList( "Eins", "Zwei" ).add( "Drei" );
```

3.2.9 测试：查找和未找到★

如果我们在 Java 程序中编写以下内容，那么输出是什么？

```
int[] numbers1 = { 1, 2, 3 };
System.out.println( Arrays.asList( numbers1 ).contains( 1 ) );
Integer[] numbers = { 1, 2, 3 };
System.out.println( Arrays.asList( numbers ).contains( 1 ) );
System.out.println( Arrays.asList( 1, 2, 3 ).contains( 1 ) );
```

3.2.10 加上奶酪，一切都会更美味：在列表中添加元素★

CiaoCiao 船长喜欢吃蔬菜，但是吃的时候必须加很多奶酪。

任务：

- 编写方法 insertCheeseAroundVegetable(List)，获得一个菜谱成分列表，只要蔬菜出现在列表中，就直接在它前面或后面添加成分"奶酪"。
- 列表必须是可修改的。

举例：

- 面团、西葫芦、辣椒、奶油、肉汤、牛奶、黄油、洋葱、番茄、盐、胡椒 → 面团、西葫芦、奶酪、辣椒、奶酪、奶油、肉汤、牛奶、洋葱、奶酪、番茄、奶酪、盐、胡椒；
- 奶酪 → 奶酪。

使用固定数量的蔬菜种类。

3.2.11 测试：一无所有让人很恼火★

给定以下代码，它应该删除列表中的所有空字符串。代码会这样做吗？

```
List<String> names = new ArrayList<>();
```

```
Collections.addAll( names, "", "Sonny", "Crockett", "Burnett",
                    "Ricardo", "", "Rico", "Tubbs", "Ricardo", "Cooper", "" );

for ( String name : names )
  if ( "".equals( name ) )
    names.remove( name );
```

3.2.12　使用迭代器搜索元素，找到 Covid Cough ★★

Bonny Brain 跑到港口寻找 Covid Cough，他将消毒剂藏在其船上。每艘船都包含一份乘客姓名列表。船舶由以下小类声明：

```
class Ship {
  private List<String> persons = new ArrayList<>();
  void addName( String name ) { persons.add( name ); }
  boolean contains( String name ) { return persons.contains( name ); }
  @Override public String toString() {
    return "" + persons;
  }
}
```

港口中有 100 艘船，它们存储在 LinkedList<Ship> 中。Covid Cough 躲在一艘不知名的船上，让我们模拟一下：

```
List<Ship> ships = new LinkedList<>();
for ( int i = 0;i< 100; i++ )
  ships.add( new Ship() );

ships.get( new Random().nextInt( ships.size() ) ).addName( "Covid Cough" );
```

Bonny Brain 到达港口的众多入口之一，左、右两边都有船只：

```
int index = new Random().nextInt( ships.size() );
ListIterator<Ship> iterator = ships.listIterator( index );
```

对船只的唯一访问方法是通过 ListIterator。注意 ListIterator 只能前进和后退，不能随机访问！

任务：
- 使用 ListIterator 访问船只，找到 Covid Cough。
- 是否有关于如何尽快找到此人的策略？已知总共有多少艘船，也就是 100 艘。既然 Bonny Brain 进入港口的索引是已知的，那么我们也知道入口左、右各有多少艘船。

3.2.13　移动元素，玩"抢椅子"游戏 ★

在生日聚会上，客人们玩起了"抢椅子"游戏（英语：musical chairs）。客人们坐在椅子上，当音乐响起时，他们起身围着椅子边走。客人的姓名在一个列表中。

任务 A：
- 创建新类 MusicalChairs。
- 创建一个在内部存储姓名的 MusicalChairs(String... names) 构造函数。
- 实现 toString( ) 方法，该方法返回以逗号分隔的姓名。
- 编写方法 rotate(int distance) 将列表中的姓名向右移动位置 distance。落在右边的元素被推回到左边。操作到位，因此（内部）列表本身发生了变化，并且该方法不返回任何内容。

> 小提示：
> 使用 Collections 中合适的方法完成这项任务。

举例：
```
MusicalChairs musicalChairs = new MusicalChairs( "Laser", "Milka", "Popo",
    "Despot" );
musicalChairs.rotate( 2 );
System.out.println( musicalChairs ); // Popo, Despot, Laser, Milka
```

任务 B：
- 编写另一个方法 void rotateAndRemoveLast(int distance)，首先将列表向右移动 distance 位置，然后删除最后一个元素。
- 添加一个 String play( ) 方法，在循环中调用 rotateAndRemoveLast(…) 直到列表中只剩下一个元素；然后确定获胜者并将其作为字符串返回。每次运行的距离是随机的。

求解时，考虑列表可能为空的情况。

3.2.14 编辑行星的问答游戏 ★★

CiaoCiao 船长在招募新船员，为了测试他们的知识，他向新船员们询问了太阳系中各行星的直径。他想要一个交互式应用程序，可以随机选择一个行星，新船员需要知道其直径。

行星被预定义为枚举类型：

```
enum Planet {

  JUPITER( "Jupiter", 139_822 ), SATURN( "Saturn", 116_464 ),
  URANUS( "Uranus", 50_724 ), NEPTUNE( "Neptune", 49_248 ),
  EARTH( "Earth", 12_756 ), VENUS( "Venus,", 12_104 ),
  MARS( "Mars", 6_780 ), MERCURY( "Mercury", 4_780 ),
  PLUTO( "Pluto", 2_400 );

  public final String name;
  public final int    diameter; // km

  Planet( String name, int diameter ) {
    this.name     = name;
    this.diameter = diameter;
  }
}
```

列表 3.1 com/tutego/exercise/util/PlanetQuiz.java

- 编写一个控制台应用程序，第一步创建所有行星的随机顺序。思考：我们如何使用 java.util.Collections 中的 shuffle(...) 方法？
- 遍历这个随机的行星序列并生成控制台输出，询问这些行星的直径。作为选项，应向新船员显示以千米为单位的四个直径，其中一个直径是正确的，三个直径来自其他行星。
- 如果新船员输入正确的直径，则屏幕上会出现一条消息；如果新船员输入了错误的直径，则控制台输出会提示正确的直径。

举例：
```
What is the diameter of planet Uranus (in km)?
49248 km
```

50724 km

12756 km

139822 km

50724

Correct!

What is the diameter of planet Pluto (in km)?

12104 km

4780 km

2400 km

12756 km

11111

Wrong! The diameter of Pluto is 2400 km.

What is the diameter of planet Jupiter (in km)?

139822 km

6780 km

2400 km

49248 km

...

3.3　集合

集合只包含它们的元素一次。它们可以是未排序的，也可以是已排序的。Java 提供了用于抽象的 Set 接口；两个重要的实现是 HashSet 和 TreeSet。

关于集合，存在一系列问题：

1. 集合为空吗？
2. 集合中有哪些元素？
3. 询问的元素是否在集合内？回答是或者不是。
4. 给出两个集合，它们包含相同的元素吗？
5. 如果两个集合合并，那么新集合是什么样的？
6. 集合是否完全包含另一个集合，即一个集合是另一个集合的子集？
7. 什么是两个集合的交集，即在两个集合中都出现的元素是什么？
8. 如果我们从一个集合中删除另一个集合中也有的元素，那么差异集是什么样的？

其中一些操作可以通过 Set 数据类型直接回答，例如方法 isEmpty( ) 或 contains(...)。尤其是集合操作的说明不是特别明白，程序员有时不得不走弯路。例如，关于子集，有一个 Collections 方法 disjoint(Collection<?>, Collection<?>)，但它返回一个 boolean 值，指示两个集合是否没有共同的元素。

我们通过任务来回答一些问题。

3.3.1 形成子集，寻找共同点 ★

Bonny Brain 的女儿正在和科拉·科罗娜约会，他们想知道他们是否般配。因此，两人都写下他们喜欢的东西。看起来是这样的：

```
Set<String> me = new HashSet<>();
Collections.addAll( me, "Candy making", "Billiards", "Fishkeeping",
    "Eating", "Action figures", "Birdwatching", "Axe throwing" );
Set<String> she = new HashSet<>();
Collections.addAll( she, "Axe throwing", "Candy making", "Action figures",
    "Casemodding", "Skiing", "Satellite watching" );
```

任务：
双方的重合度是百分之多少？我们可以用什么方法回答这个问题？

小提示：
查看 Set 中的方法，它们是否能够形成子集或交集？

3.3.2 测试：好剑 ★

编译以下程序。程序的输出是什么？

```
class Sword implements Comparable<Sword> {
  String name;

  public int compareTo( Sword other ) { return 0; }

  public boolean equals( Object other ) {
    throw new IllegalStateException();
  }

  public String toString() { return name; }

  public static void main( String[] args ) {
    Sword one = new Sword();
    Sword two = new Sword();
    one.name = "Khanda";
    two.name = "Kilij";
    Set<Sword> swords = new TreeSet<>();
    System.out.printf( "%s%s%s", swords.add( one ), swords.add( two ),
      swords );
  }
}
```

3.3.3 删除数组中的重复元素★

任务：

- 创建静态方法 double[] unique(double... values)，它返回一个数组，其中传入的数组的所有元素都以相同的顺序排列，但没有重复的条目。
- 如果参数为 null，则该方法必须抛出异常。

举例：

```
System.out.println( Arrays.toString( unique() ) );
// []
System.out.println( Arrays.toString( unique( 1, 2 ) ) );
// [1.0, 2.0]
System.out.println( Arrays.toString( unique( 1, 1 ) ) );
// [1.0]
System.out.println( Arrays.toString( unique( 1, 2, 1 ) ) );
```

```
                              // [1.0, 2.0]
System.out.println( Arrays.toString( unique( 1, 2, 1, Double.NaN ) ) );
                              // [1.0, 2.0, NaN]
System.out.println( Arrays.toString( unique( 1, Double.NaN, Double.
NaN ) ) );
                              // [1.0, NaN]
System.out.println( Arrays.toString( unique( -0, 0 ) ) );
                              // [1.0, 2.0, NaN]
```

列表 3.2 com/tutego/exercise/util/UniqueArrayElements.java

3.3.4　查明单词中包含的所有单词★★

CiaoCiao 船长截获了一条秘密信息，文本由看似无关的单词组成。他想了想才发现，文本里还藏着别的单词。在单词 Rhabarbermarmelade 中包含单词 Rhabarber, marmelade，arme，arm，lade，hab。

程序应该找出给定单词包含哪些有效单词。要确定有效单词是什么，我们可以使用网上的单词列表：

- ▶ https://raw.githubusercontent.com/dwyl/english-words/master/words_alpha.txt（英语）
- ▶ https://raw.githubusercontent.com/creativecouple/all-the-german-words/master/corpus/de.txt（德语）

任务：

使用静态方法 Collection<String> wordList(String string, Collection<String> words) 编写一个程序，该方法生成 string 中包含的所有子字符串，并准确返回 Collection 中那些在字典 words 中有效的单词。

英语字典示例：

- ▶ wordList("wristwatches", words) → [wrist, wristwatch, wristwatches, rist, ist, twa, twat, wat, watch, watches, tch, tche, che, hes]
- ▶ bibliophobia → [abib, bib, bibl, bibliophobia, pho, phobia, hob, obi, obia]

小提示：
可以将带有单词的文件转换为这样的数据结构：
```
private static final String WORD_LIST_URL =
    "https://raw.githubusercontent.com/creativecouple/all-the-german-
```

```
words/master/corpus/de.txt";
    // "https://raw.githubusercontent.com/dwyl/english-words/master/
    // words_alpha.txt";
private static Collection<String> readWords() throws IOException {
  URL url = new URL( WORD_LIST_URL ); // 370.000 words
  Collection<String> words = new HashSet<>( 500_000 );
  try ( InputStream is = url.openStream() ) {
    new Scanner( is ).forEachRemaining( s -> words.add( s.toLowerCase() ) );
  }
  return words;
}
```

3.3.5 正确分类几乎相同的东西★★

给出一个字符串的集合，类似这样：

"-13.123", "0", "0", "10101010", "10101010.0", "0.0", "-0.0"

任务：
- 列表将被排序，重复的元素将被删除，例如第二个"0"。
- 虽然"0""0.0""-0.0"数值相同，但仍要保留成三个条目。
- 数字可以具有任意长度和任意数量的小数位。

如果我们要输出排序后的上方列表，则屏幕上会出现以下内容：

[-13.123, -0.0, 0, 0.0, 10101010, 10101010.0]

3.3.6 用 UniqueIterator 排除重复元素★★

在其他数据结构中，例如列表，元素可以出现不止一次。编写一个 UniqueIterator，只返回一个 Collection 的元素一次。

通用类型的声明如下：

```
public class UniqueIterator<E> implements Iterator<E> {
```

```
    public UniqueIterator( Iterator<? extends E> iterator ) {
      // ...
    }

    // usw.
}
```

构造函数显示，新的迭代器得到一个现有的迭代器作为参数。因此，调用可能看起来像这样：

```
List<String> names = ...;
Iterator<String> UniqueIterator = new UniqueIterator( names.iterator() );
```

3.4 关联映射

关联映射将键与值关联。它们在其他编程语言中也称为字典。Java 通过 Map 接口规定了所有实现的操作。两个重要的实现是 HashMap 和 TreeMap。

3.4.1 将二维数组转换为映射 ★

从 Collection 继承的数据类型在接受数据方面相对灵活，例如，可以使用 addAll(Collection) 将 List 的元素复制到 Set 中。数组也可以通过 Arrays.asList(...) 直接用作 Collection。

Map 数据类型不太灵活，将数组或其他 Collection 集合转换为 Map 并不容易。

任务：
- 编写方法 Map<String, String> convertToMap(String[ ][ ])，将二维数组转换为 java.util.Map。
- 数组中的第一个条目应该是键，第二个条目应该是值。
- 键正确实现 hashCode( ) 和 equals(...)。
- 如果稍后在数组中再次出现相同的键，则该方法将覆盖之前的键值对。
- 键和值不能为 null，否则必须抛出异常。

举例：
```
String[][] array = {
    { "red", "#FF0000" },
    { "green", "#00FF00" },
```

```
        { "blue", "#0000FF" }
};
Map<String, String> colorMap = convertToMap( array );
System.out.println( colorMap ); // {red=#FF0000, green=#00FF00, blue=#0000FF}
```

3.4.2 将文本转换为摩尔斯密码并反转 ★

CiaoCiao 船长必须通过摩尔斯密码向一个遥远的岛屿发送信息。摩尔斯密码由长、短符号组成，用字符"."和"-"表示。

把下面的定义复制到新类 Morse 中：

```
// A .-      N -.       0 -----
// B -...    O ---      1 .----
// C -.-.    P .--.     2 ..---
// D -..     Q --.-     3 ...--
// E .       R .-.      4 ....-
// F ..-.    S ...      5 .....
// G --.     T -        6 -....
// H ....    U ..-      7 --...
// I ..      V ...-     8 ---..
// J .---    W .--      9 ----.
// K -.-     X -..-
// L .-..    Y -.--
// M --      Z --..
```

任务：

编写两个方法。

- String encode(String string)。它接收一个字符串并将其转换为摩尔斯密码。字符串的每个字符都应该以相应的摩尔斯密码输出。每个代码块应在输出中用空格分隔。未知字符被跳过。小写字母应像大写字母一样被处理。单词之间有两个空格。

- String decode(String string)。将摩尔斯密码转换回原始字符串。单词分隔符的两个空格再次变为单个空格。

3.4.3 用关联映射标记词频 ★★

绯闻女孩在窃听甲板上的一群人，这样稍后她就可以告诉 CiaoCiao 船长这些人正在讨论什么。重要的是什么词或词组经常出现。

任务：
编写方法 List<String> importantGossip(String...words)，从字符串的可变参数中准确返回一个包含 5 个字符串的列表，其在传递的数组中出现频率最高。

举例：
```
String[] words = {
    "Baby Shark", "Corona", "Baby Yoda", "Corona", "Baby Yoda", "Tiger King",
    "David Bowie", "Kylie Jenner", "Kardashian", "Love Island", "Bachelorette",
    "Baby Yoda", "Tiger King", "Billie Eilish", "Corona"
};
System.out.println( importantGossip( words ) );
```

输出如下：
[Baby Yoda, Corona, Tiger King, Baby Shark, Bachelorette]

牢记，搜索的不是 Baby 或 Yoda 这样的单词，而是整个字符串，如 Baby Yoda 或 Baby Shark。

3.4.4 读取颜色并播放★★

Bonny Brain 拿到了她的旗帜的新设计，但设计师说的话让人难以理解：

Für den Hintergrund nehmen wir #89cff0 oder #bcd4e6 und für den Text vielleicht #fff af0 oder #f8f8ff.

（For the background we'll use #89cff0 or #bcd4e6 and for the text maybe #fffaf0 or #f8f8ff.）

（背景我们使用 #89cff0 或者 #bcd4e6，文字使用 #fffaf0 或 #f8f8ff。）

她发现，像 #RRGGBB 这样的规范代表了颜色的红、绿、蓝部分，以十六进制编码。幸运的是，有一些"翻译表"（如 https://tutego.de/download/colors.csv），其中的行包含诸如以下内容：

```
amber,"Amber",#ffbf00,255,191,0
aqua,"Aqua",#0ff,0,255,255
blush,"Blush",#de5d83,222,93,131
```

```
wine,"Wine",#722f37,114,47,55
```

在某些情况下，颜色值仅在文件中带有三个符号，例如示例中带有 0ff 的 aqua。在这种情况下，单个颜色规格加倍，因此 #RGB 变为 #RRGGBB。

任务：

▶ 为颜色的表示创建一个新类 Color。每种颜色都有一个名称 (String name) 和一个 RGB 值（int rgb）。编写（或从 IDE 生成）toString( ) 方法。如果需要，添加其他方法。

▶ 创建新类 ColorNames。
  - 给类一个对象变量 HashMap<Integer, Color> colorMap 以便 ColorNames 可以在内部记住一个映射中的所有 Color 对象；map 的键是整数 RGB 值，关联的值是对应的 Color 对象。
  - 将文件 https://tutego.de/download/colors.csv 复制到本地硬盘中。
  - 创建一个读取文件的构造函数。我们可以使用 Scanner 来执行此操作，或者使用返回 List<String> 的 Files.readAllLines(Paths.get("colors.csv")) 读取整个文件。
  - 拆分 CSV 源的每一行，提取颜色名称（第 2 列）和 RGB 值（第 3 列）。提示：可以使用 Java 方法将颜色值转换为整数——Integer.decode("#722f37") 返回 7483191。请注意，颜色规范可以以 #RGB 和 #RRGBB 的形式出现。
  - 将颜色名称和整数值传输到 Color 对象并将它们放入 Map。
  - 添加方法 decode(int rgb)，返回一个 RGB 值的关联 Color 对象。

举例：

▶ mapper.decode(7483191) → Optional['Wine' is RGB #722F37];
▶ mapper.decode(7) → Optional.empty。

3.4.5 读取名称，管理长度★★

Bonny Brain 喜欢玩名字填字游戏，每个条目都是一个名字。她经常想不出一个有一定长度的名字——她需要软件的帮助！

任务：

1. 文件 http://tutego.de/download/family-names.txt 包含姓氏。将文件保存在自己的文件系统中。
2. 读取文件。例如，我们可以使用 Scanner 类或 Files 方法 readAllLines(Path)。
3. 将名字排序成 TreeMap<Integer, List<String>>：关键是名字的长度，列表包含所有长度相同的名字。
4. 按长度升序列出命令行中的所有名字。
5. 从命令行询问长度并输出该长度的所有名字，询问的长度不为 0 或负数。

3.4.6　找到缺失的字符★★

CiaoCiao 船长偷走了死海古卷（Cumexhopp-Rollen），至今没有人能够破译其中的文字。他想做到！然而，许多字符难以辨认，虽然其他字符是清晰的。不过可以看出一个单词有多少个字符。

任务：
- 使用 main(...) 方法创建一个新类并将两个列表复制到程序中：
  ```
  List<String> words = Arrays.asList( "haus", "maus", "elefant", "klein",
      "groß" );
  List<String> missingLettersWords = Arrays.asList( "ha__", "el__a_t", "x",
      "hi__", "___s" );
  ```
- 将 missingLettersWords 中的每个单词与字典中所有可能的单词进行匹配，其中下划线象征未知字符。
- 词典中建议的单词的长度必须等于"难以辨认"的单词的长度。
- 必须至少有一个字符。

举例：
给定列表的输出可能如下：

```
ha__ -> [haus]
el__a_t -> [elefant]
x -> No results
hi__ -> No results
___s -> [maus, haus]
___s___ -> No results
```

3.4.7　计算寻找三头猴的路径数★★

在曼哈顿喝了一夜酒后，CiaoCiao 船长弄丢了他的三头猴。他一定是在路上的某个地方把毛绒玩具弄丢了！但它会在哪里呢？他的团队必须从头到尾走完所有的街道。CiaoCiao 船长唯一能记得的是，他没有走对角线。

图 3.2 中显示的是 4×4 的街区，有 14 种可能。经过一段时间的寻找船员们幸运地找到了毛绒玩具！

CiaoCiao 船长在想：如果有 5 个或 10 个街区，会发生什么情况？路径的数量岂不是太多，难以搜索？

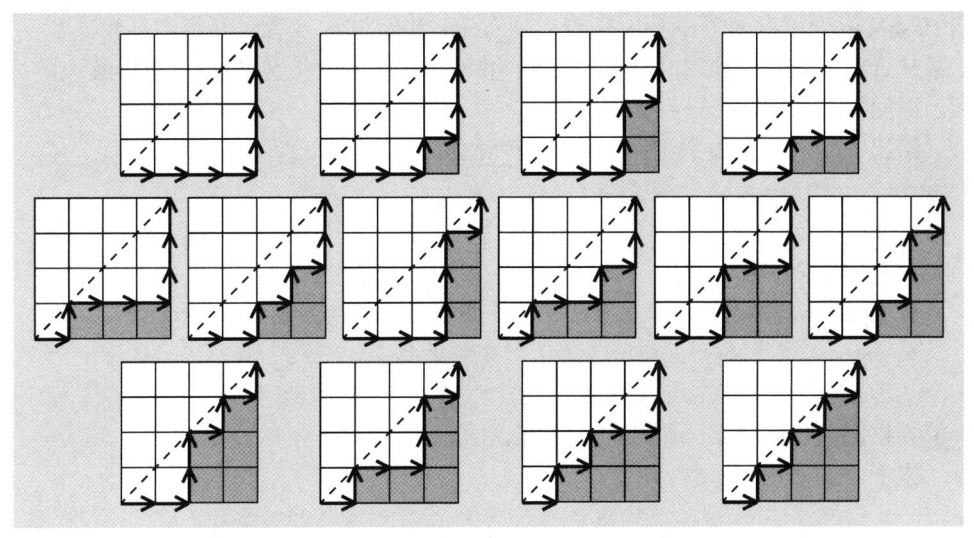

图 3.2　从头到尾的可能路径（来源：维基百科）

数学提供了问题的答案。这里寻求的是具有 n×n 个单元格的正方形的单调路径。加泰罗尼亚数字提供了可能的路径数，其计算方法如下：

Cn = (2n)! / (n+1)! n!

任务：
- 通过 BigInteger catalan(BigInteger n) 方法转换公式。访问内部的 BigInteger factorial(BigInteger n) 方法进行阶乘计算。
- 公式中要计算三个阶乘：n!, (n+1)! 和 (2n)!。(n+1)! 就是 n!(n+1)，也就是 n! 要计算两次；计算 (2n)! 时也出现了中间结果 (n+1)!。许多乘法都要进行两次，因此应该对乘积进行缓存。为此我们使用数据类型 WeakHashMap。
- 比较我们用相同的参数两次调用 catalan(...) 方法的时间。使用以下代码作为模板：

```
long start = System.nanoTime();
BigInteger catalan1000 = catalan( BigInteger.valueOf( 1000 ) );
long end = System.nanoTime();
System.out.println( catalan1000 );
System.out.println( TimeUnit.NANOSECONDS.toMillis( end - start ) );
```

3.4.8 在排序的关联映射中管理节日 ★

在 TreeMap 中，元素是自动排序的。TreeMap 实现了 java.util.NavigableMap，而 HashMap 则没有。顺序是由外部 Comparator 决定的，或者元素包含自然顺序。

从 API 文档我们得知，firstEntry( ) 和 lastEntry( ) 分别返回最小的和最大的元素。返回类型为 Map.Entry<K,V>。

给出一个键，下面的方法会返回一个相对于这个键的值：

- ceilingXXX(K key): 返回一个大于或等于此键的结果。
- floorXXX(K key): 返回一个小于或等于给定键的结果。
- lowerXXX(K key): 返回一个小于给定键的结果。
- higherXXX(K key): 返回一个大于给定键的结果。

如果问题没有答案，则所有方法都返回 null。
以下方法适用于子集：

- SortedMap<K,V> subMap(K fromKey, K toKey);
- NavigableMap<K,V> subMap(K fromKey, boolean fromInclusive, K toKey, boolean toInclusive);

对于第一种方法，fromKey 是包含的，toKey 是排他的，这对应 Java 的惯例。第二种方法可以更精确地控制开始或结束元素是否被包含在内。

任务：

- 创建一个排序的关联映射：
  ```
  SortedMap<LocalDate, String> dates = new TreeMap<>();
  dates.put( LocalDate.of(...), "..." );
  ```
  标准化 <LocalDate, String> 意味着 LocalDate 时间类型应与字符串关联。
- LocalDate 类实现 Comparable，这意味着元素具有自然顺序。
- 在数据结构中填写一些真实或虚构节日的配对。
- 使用适当的 NavigableMap 方法回答以下问题：
  - 根据数据结构，最早和最后一个节日是什么时候？
  - 圣诞假期于 01.06 结束。圣诞假期后的第一个节日是什么时候？
  - 12.23—01.06 的圣诞假期包括哪些日期值（日期值包含在内）？
  - 从数据结构中删除圣诞假期。

3.4.9 测试：HashMap 中的值 ★★

思考：以下程序的屏幕输出是什么？

```
Map<Point, String> map = new HashMap<>();
Point p = new Point( 1, 2 );
map.put( p, p.toString() );
p.setLocation( 2, 1 );
System.out.println( map );
System.out.println( map.get( p ) );
p.setLocation( 1, 2 );
System.out.println( map.get( p ) );
```

对于关联映射的值，什么是必需的？

3.4.10 确定共同点：派对场所布置和带来的礼物 ★

Bonny Brain 计划举办一个派对，所有家庭都会带一些礼物来：

```
Set<String> gombonoGifts = new HashSet<>();
Collections.addAll( gombonoGifts, "Vodka", "BBQ Grill", "kneading soap" );

Set<String> banannaGifts = new HashSet<>();
Collections.addAll( banannaGifts, "Vodka", "drinking helmet" );

Set<String> cilimbiGifts = new HashSet<>();
Collections.addAll( cilimbiGifts, "drinking helmet", "Money box", "Vodka",
    "water pistol" );

List<Set<String>> families = Arrays.asList( gombonoGifts, banannaGifts,
    cilimbiGifts );
```

Bonny Brain 是一个完美主义的战略家，她想知道是否有礼物被重复带来。

任务：

▶ 编写方法 printMultipleGifts(List<Set<String>> families)，输出某个礼物被带来多少次。

▶ 什么礼物不止一次被带过来？

举例：
以上赋值的输出可能如下：

```
{drinking helmet=2, kneading soap=1, water pistol=1, Money box=1, BBQ Grill=1, Vodka=3}
drinking helmet
Vodka
```

3.5　Properties

　　Properties 类是一个特殊的关联映射，只将字符串与字符串关联。该类不仅代表一个数据结构，还可以读写文件，即所谓的属性文件。这些文件是文本文件，通常用于配置。键值对在文件中以"="分隔。这些值也可以用 XML 格式读写，但不常见。

3.5.1　开发便捷的属性装饰器★★

　　Properties 类包含键值对，其中的键始终是字符串。可能的转换必须由开发人员自己完成，比较麻烦。

任务：
编写类 PropertiesConfiguration，用于装饰一个 Properties 对象。如果该键不存在，则最普通的方法是返回 Optional，它要么是已填充的，要么是空的。

▶ Optional<String> getString(String key)

　　Optional 的优点是可以简单地确定默认值的替代：conf.getProperty("rank").orElse("Captain")。

PropertiesConfiguration 的其他方法应进行转换：

▶ Optional<Boolean> getBoolean(String key);
▶ OptionalLong getLong(String key);
▶ OptionalDouble getDouble(String key);
▶ Optional<BigInteger> getBigInteger(String key);

如果没有该键的关联值，则该容器为空。向错误的转换失败也会导致一个空的容器。

API 的示例：

```
Properties root = new Properties();
root.setProperty( "likes-rum", "true" );
root.setProperty( "age", "55" );
root.setProperty( "income", "123456789012" );
root.setProperty( "hobbies",
   "drinking, gambling\\, games, swearing competitions" );
root.setProperty( "weakness_of_character", "" );
PropertiesConfiguration conf = new PropertiesConfiguration( root );
Optional<Boolean> maybeLikesRum = conf.getBoolean( "likes-rum" );
OptionalLong maybeAge = conf.getLong( "age" );
Optional<BigInteger> maybeIncome = conf.getBigInteger( "income" );

System.out.println( maybeLikesRum );   // Optional[true]
System.out.println( maybeAge );        // OptionalLong[55]
System.out.println( maybeIncome );     // Optional[123456789012]
```

可选补充：询问列表

高级开发人员可以实现以下方法：

- List<String> getList(String key)。返回一个以逗号分隔的字符串。逗号本身可以用"\"掩盖。

举例：

```
List<String> hobbies = conf.getList( "hobbies" );
List<String> weaknessOfCharacter = conf.getList( "weakness_of_character" );

System.out.println( hobbies );                    // [drinking, gambling, games,
                                                  // swearing competitions]
System.out.println( hobbies.size() );             // 3
System.out.println( weaknessOfCharacter );        // []
```

可选补充：存储二进制值

java.util.HashMap 可以关联任何类型，Properties 只能将字符串与字符串关联。如果要存储其他数据类型，例如 byte 数组，则必须将它们转换为字符串。byte[] 可以通过多种方式转换为 ASCII 字符串，包括 BASE64 编码；Java 可以通过 Base64 类做到这一点。

由于 Properties 是读取而不是写入的，所以到目前为止，getXXX(...) 方法对我们来说已经足够了。在下面的补充中，我们要编写两个新方法，一个用于设置，一个用于查询：

- void putBinary(String key, byte[] bytes);
- Optional<byte[]> getBinary(String key)。

应用示例：
```
conf.putBinary( "binary", new byte[]{ 0, 1, 127, (byte) 254, (byte) 255 } );
System.out.println( conf.getString( "binary" ) ); // Optional[AAF//v8=]
byte[] binary = conf.getBinary( "binary" ).get();
System.out.printf( "%d%d%d%d%d", binary[0], binary[1], binary[2],
  binary[3], binary[4] );
```

3.6 堆栈 (Stack) 和队列 (Queue)

在 Java 中，通用列表允许通过索引访问元素，也称这种访问为随机访问，因为我们可以选择在任意位置查询元素。有些数据结构受到明显更多的限制，例如只能在开头或结尾插入或删除元素。

其中包括：

- 堆栈 (Stacks)；
- 队列 (Queues)。

使用堆栈时，我们只能在一端插入元素，并且必须在该端移除元素。该原则也称为"后进先出"（Last In，First Out，LIFO）。与此相反的是队列。它先读出的是先添加的内容。该原则叫作"先进先出"（First In，First Out，FIFO）。

Java 中没有纯粹的堆栈和队列，只有由列表实现的接口。

3.6.1 编辑 RPN（逆波兰表示法）计算器 ★

我们通常用中缀表示法写数学表达式，其中运算符在操作数之间，例如 47 +

11，但是原则上，运算符也可以在操作数的前面，例如 + 47 11，或在它们后面，例如 47 11 +。

在 20 世纪 80 年代，惠普计算机建立了一种称为逆波兰表示法 (RPN) 的特殊输入法。这是一种后缀表示法，运算符跟在值后面。对计算机来说其优势在于用户已经确定了优先级——点计算在线计算之前，这简化了计算器中的程序逻辑。

PostScript 也使用这种表示，因为可以使用堆栈轻松解析数学表达式。

我们想编写一个 RPN 计算器程序。

任务：
- 编写一个程序，首先将"12 34 23 + *"这样的字符串拆分成若干个标记。
  提示：要拆分一个字符串，我们可以使用 String 的 split(...) 或 Scanner。
- 拆分字符串后应提取结果。从一个固定的字符串开始，进行测试。
- 从命令行读取一个字符串，这样我们就有一个真正的 RPN 计算器。
- 哪些错误和问题需要处理和拦截？我们应该如何处理错误？

3.7 BitSet

BitSet 类是 boolean 数组的一种节省空间的高性能替代方案。当我们需要将整数映射到逻辑值时，该数据结构很有用。数据结构可以快速回答一个索引（一个正整数）是与 true 关联还是与 false 关联。如果位数太多或有很大差距，那么 https://github.com/brett wooldridge/SparseBitSet 是一个不错的选择。

3.7.1 查找重复条目并解决动物混乱 ★

CiaoCiao 船长在临睡前给他的私人动物园的动物们喂食，但由于他喝了朗姆酒后有点醉，所以忘了关大门。第二天早上，加比·格拉特和弗雷德·弗里特注意到，动物们都不见了。他们迅速跑到 CiaoCiao 船长那里报告："有些动物逃跑了！" CiaoCiao 船长问道："天哪！哪些动物？"两人边想边画（写字不是他们的强项）：

- 🐗🐘🦋🐝
- 🐒🐘🐅🐸🦋🐋

CiaoCiao 船长发现，这两人记性很差，因此只想寻找他们俩都提到的动物。

任务：
- 编写方法 String sameSymbols(String, String)，该方法返回一个包含共同符号

的字符串。顺序无关紧要，所有 Unicode 字符都可用。
- 由于我们需要遍历 String 并且它包含两个 char 之外的"更高级"的 Unicode 字符，所以解决方案应该使用 string.codePoints( ).forEach(consumer)。此语句遍历字符串 string 的所有字符并为每个字符调用提供的 IntConsumer。这是 Stream API 的一个应用程序，我们将在下一章中更详细地介绍它。

举例：
- sameSymbols("🐑🐖🦋🐛", "🐙🐒🐧🐸🐛🐙") → "🐑🦋";
- sameSymbols("abcy", "bcd") → "bc";
- sameSymbols("abc", "def") → ""。

由于着急要结果，所以该方法的实现方式是运行时间与字符串的长度呈线性关系，用计算机术语来说：如果 N 和 M 是字符的长度，则运行时间为 $O(N+M)$。允许使用所有 Unicode 字符。

3.8 线程安全的数据结构

对于之前关于数据结构 ArrayList，LinkedList，HashSet，TreeSet 等的任务，我们只需要 Main 线程即可。下面的任务涉及更多的线程和并发访问，那么我们就需要使用线程安全的数据结构。对于解决方案，我们使用 java.util.concurrent 包中的数据类型，其中声明了一系列线程安全的数据结构，它们即使在任意数量的并行访问下也能正常工作，并且具有非常好的性能。

3.8.1 装船★★

CiaoCiao 船长和船员们正在为加佐帕佐普岛上的下一次伟大冒险做准备。5 名船员将箱子和桶放在装载坡道上，10 名船员将货物存放在船上。最多可同时在装载坡道上放置 5 个对象。

任务：
- 为装载坡道创建容量为 5 的 ArrayBlockingQueue<String>。
- 创建两个不同的 Runnable 实现 Loader 和 Unloader 以获取对 ArrayBlockingQueue 的引用。
  - Loader 将字符串放在装载坡道上（来自一组对象名称的随机字符串）。
  - Unloader 应从装载坡道上获取字符串并在屏幕上输出。
  Loader 和 Unloader 的工作随机需要 1~2 秒。
- Unloader 有 5 个线程，Loader 有 10 个线程。

小提示：
针对添加和删除有不同的方法。区分很重要，否则可能出现程序错误，见表 3.1。

表 3.1 用于添加和删除元素的 BlockingQueue 方法

操作	异常	null 返回	阻塞
添加	add(e)	offer(e)	put(e)
删除	remove( )	poll( )	take( )

我们不能从方法名称中派生语义，必须学习其中的区别。只有一列和方法适合我们的任务。

3.8.2 优先编辑重要消息 ★

PriorityQueue 具有内部排序功能，以便具有更高优先级的元素可以移到前面。优先级来自实现 Comparable 的元素的自然顺序或外部 Comparator。"小"元素有更高的优先级并移到 PriorityQueue 的前面。在队列的末尾是具有最低优先级的元素（这与疫苗接种相同：优先级组 1 首先获得疫苗）。

CiaoCiao 船长被分配了来自各方面的任务。当然，Bonny Brain 的需求永远是第一位的。CiaoCiao 船长能识别他的任务，这是因为 Bonny Brain 的工作委托中包含昵称"Kanönchen"。

任务：

▶ 编写类 Message，存储用于工作委托的 String 和类型 long 的时间戳的消息。使用 System.nanoTime( ) 初始化时间戳。实现 / 生成 hashcode( )、equals(Object) 和 toString( )。

▶ 为 Message 实现一个 Comparator，它创建一个顺序，以便有昵称的消息比没有昵称的消息"更小"，稍后可以将其视为更高的优先级。如果两个消息都包含昵称或两个消息都不包含昵称，则它们具有相同的"大小"。

▶ 使用另一个比较逻辑扩展 Comparator，以便将时间戳考虑在内，更早的消息可以更早地被处理。

▶ 使用消息初始化 PriorityQueue 并观察有昵称的消息在队列中向前移动。

举例：

假设 PriorityQueue<Message> tasks 是正确初始化的数据结构，以下程序将产生如下所示的输出：

```
tasks.add( new Message( "Treasure Hunt" ) );
System.out.println( tasks );

tasks.add( new Message( "Kanönchen, Family Movie Night!" ) );
System.out.println( tasks );

tasks.add( new Message( "Build a pirate ship" ) );
System.out.println( tasks );

System.out.println( tasks.remove() );
System.out.println( tasks );

System.out.println( tasks.remove() );
System.out.println( tasks );

tasks.add( new Message( "Capture the Flag" ) );
System.out.println( tasks );

tasks.add( new Message( "Bury the treasure, Kanönchen" ) );
System.out.println( tasks );

tasks.add( new Message( "Kanönchen, make a treasure map" ) );
System.out.println( tasks );

System.out.println( tasks.remove() );
System.out.println( tasks );

System.out.println( tasks.remove() );
System.out.println( tasks );

System.out.println( tasks.remove() );
System.out.println( tasks );
```

```
System.out.println( tasks.remove() );
System.out.println( tasks );
```

输出如下：

['Treasure Hunt', 46400]

['Kanönchen, Family Movie Night!', 23700, 'Treasure Hunt', 46400]

['Kanönchen, Family Movie Night!', 23700, 'Treasure Hunt', 46400, 'Build a pirate ship', 70600]

'Kanönchen, Family Movie Night!', 23700

['Treasure Hunt', 46400, 'Build a pirate ship', 70600]

'Treasure Hunt', 46400

['Build a pirate ship', 70600]

['Build a pirate ship', 70600, 'Capture the Flag', 83200]

['Bury the treasure, Kanönchen', 10900, 'Capture the Flag', 83200, 'Build a pirate ship', 70600]

['Bury the treasure, Kanönchen', 10900, 'Kanönchen, make a treasure map', 71400, 'Build a pirate ship', 70600, 'Capture the Flag', 83200]

'Bury the treasure, Kanönchen', 10900

['Kanönchen, make a treasure map', 71400, 'Capture the Flag', 83200, 'Build a pirate ship', 70600]

'Kanönchen, make a treasure map', 71400

['Build a pirate ship', 70600, 'Capture the Flag', 83200]

'Build a pirate ship', 70600

['Capture the Flag', 83200]

'Capture the Flag', 83200

[]

3.8.3 用完就换新的 ★★★

表达式 new BigInteger(1024, new SecureRandom( )) 产生一个 BigInteger 类型的随机大数。

任务：

▶ 编写自己的 SecureRandomBigIntegerIterator 类，该类实现 Iterator 并可以返回无限数量的 BigInteger。

▶ 每当数字被查询并"用完"时，后台线程应该自动计算一个新的随机数。

3.9 建议解决方案

测试 3.1.1：搜索 StringBuilder
程序返回输出：

```
true
false
```

对此的解释在于方法 contains(Object) 的实现。它在内部使用返回元素位置的方法，但我们可以从 Java 库的当前实现中读取必须应用于搜索的内容：

```
int indexOfRange(Object o, int start, int end) {
    Object[] es = elementData;
    if (o == null) {
        for (int i = start; i < end; i++) {
            if (es[i] == null) {
                return i;
            }
        }
    } else {
        for (int i = start; i < end; i++) {
            if (o.equals(es[i])) {
                return i;
            }
        }
    }
    return -1;
}
```
列表 3.3 Ausschnitt aus java.lang.ArrayList

indexOfRange(...) 方法搜索对象 o 的位置，首先区分是否必须搜索 null 引用还是常规对象。在我们的例子中，请求中的对象不为 null，因此使用 equals(...) 方法，即循环通过 equals(...) 方法将列表中的每个元素与我们的值进行比较。相反，这意味着查询只有在我们也实现了合理的 equals(...) 方法时才有效。这正是问题所在：String 类提供了一个 equals(...) 方法，不是 java.lang.Object 中的那个，而是一个被重写的方法。但是，StringBuilder 类没有实现 equals(...) 方法，而是从超类 Object 继承了该方法。然而，在超类中只实现了引用比较，这意味着内容根本无关紧要。由于在列表中的 StringBuilder 对象总是不同的，所以引用比较永远不会得到 true。

任务 3.2.1：唱歌做饭：运行列表，检查特性

存在不同的方法可以完成任务。一种变体是在一个变量中记录厨师的数量，在另一个变量中记录乐手的数量，最后比较两个变量。

建议解决方案实现了另一个变体。

```java
public static boolean areSameNumberOfCooksAndMusicians( List<CrewMember>
    crewMembers ) {
  int weight = 0;
  for ( CrewMember member : crewMembers ) {
    switch ( member.profession ) {
      case COOK: weight++; break;
      case MUSICIAN: weight--; break;
    }
  }
  return weight == 0;
}
```

列表 3.4 com/tutego/exercise/util/SameNumberOfCooksAndMusicians.java

当涉及两种东西的数量是否相等的问题时，就像有一个天平，左边和右边的重量必须相等。有趣的是，实际的重量一点也不重要，重要的是天平处于平衡状态。

此处实现的解决方案在列表中运行，如果列表中有厨师，则增加变量 weight，如果列表中有乐手，则减少变量；当然也可以反过来。如果厨师和乐手的数量相同，则最终 weight 为 0。

职业的比较通过 switch-case 实现，当然也可以使用 if 构造和条件运算符来实现。从 Java 14 开始，还有两个选项。第一个变体如下所示：

```java
switch ( member.profession ) {
  case COOK     -> weight++;
  case MUSICIAN -> weight--;
}
```

在 switch 语句的这种新写法中，用箭头代替了冒号，并且不再需要 break。从

Java 14 开始，关键字 switch 可以在另一个变体中使用，作为表达式：

```
for ( CrewMember member : crewMembers ) {
  weight += switch ( member.profession ) {
    case COOK -> +1;
    case MUSICIAN -> -1;
    default -> 0;
  };
}
```

列表 3.5 com/tutego/exercise/util/SameNumberOfCooksAndMusicians.java

在这个变体中，我们必须引入一个 default 分支。这样基本上一行会有更多代码，这种写法很可能不如第一种写法或带有条件运算符的变体那么有吸引力。

还有另一种解决方案，它不需要条件判断。基本思想仍为创建一个天平，只是将一个枚举元素转移到 +1，将另一个枚举元素转移到 −1 的方式表达起来不一样：

```
int result = 0;
for ( CrewMember member : crewMembers ) {
  //                                      CAPTAIN -+
  //                                    NAVIGATOR -+ |
  //                                    CARPENTER -+ | |
  //                                         COOK -+ | | |
  //                                     MUSICIAN -+ | | | |
  //                                               v v v v v
  int zeroOrOneOrTwo = ((1 << member.profession.ordinal()) & 0b1_1_0_0_0) / 8;
  int minusOneOrZeroOrPlusOne = (zeroOrOneOrTwo / 2) - (zeroOrOneOrTwo & 1);
  result += minusOneOrZeroOrPlusOne;
}
return result == 0;
```

列表 3.6 com/tutego/exercise/util/SameNumberOfCooksAndMusicians.java

乍一看，这种方法可能难以理解，但阅读并理解这种路径是有好处的。我们来推导解决方案。

给出一个包含各种元素的枚举，所有元素都有一个位置，即所谓的序数。CrewMember.Profession.CAPTAIN.ordinal( ) 为 0，并且 CrewMember.Profession.COOK.ordinal( ) 为 3。如果我们写 1 << x，并且 x 介于 0 和 5 之间，将 1 向左移动 x 位置

并在右侧填充0。换句话说，我们得到了二进制写法的数字：

- 0b000001 (CAPTAIN);
- 0b000010 (NAVIGATOR);
- 0b000100 (CARPENTER);
- 0b001000 (COOK);
- 0b010000 (MUSICIAN);
- 0b100000 (DOCTOR)。

0b001000 (COOK) 和 0b010000 (MUSICIAN) 是我们所关心的。为了测试数字中是否设置了第三位或第四位，我们将该数字与位模式 0b11000 连接起来。如果设置了两个位中的一个，则保留该位，将所有其他位通过与 0 的 And 连接设置为 0。最后，因为不能同时设置两个位，所以剩下 0（没有匹配）、0b10000（16）或 0b01000（8）。如果我们将该值除以 8（或将其向右移动三个位置），那么我们得到 0，1 或 2。

如果我们想再次使用天平原理，则数字 1 和 2 没有帮助。我们必须使用 1 → –1 和 2 → +1 或 1 → +1 和 2 → –1，并且 0 必须保持为 0。当然我们可以使用条件判断，但我们会尽力避免使用它。如果 x 为数字 0，1，2，那么表达式 (x / 2) – (x & 1) 转换为所需的目标，–1 和 +1。我们可以添加表达式并像之前那样做。

在正常情况下，没有人编写这样的解决方案，除非它们对性能的作用非常关键，并且分析器表明这种变体更快。因为我们在这里有一些数学运算——/8，/2 也很便捷——这最终不会比 if 快。大家喜欢尝试在微优化中重写条件判断，但如果不知道自己在做什么，最终会得不偿失。

任务 3.2.2：从列表中过滤评论

```
public static void reduceToComments( List<String> lines ) {

  if ( lines.size()% 4 != 0 )
    throw new IllegalArgumentException(
        String.format( "Illegal size%d of list, must be divisible by 4",
          lines.size() ) );

  for ( int blockStart = lines.size() - 4; blockStart >= 0; blockStart -= 4 )
  {
    // keep element at position blockStart + 3
    lines.remove( blockStart + 2 );
```

```
      lines.remove( blockStart + 1 );
      lines.remove( blockStart + 0 );
    }
  }
```

列表 3.7 com/tutego/exercise/util/RetainComments.java

要使算法正常工作，列表长度必须是 4 的倍数。因此，第一种条件判断检查长度是否能被 4 整除；如果不能，则出现异常。

以下带有索引 blockStart 的循环从前往后分四步运行。blockStart + 0，blockStart + 1，blockStart + 2 和 blockStart + 3 这四个和代表了一个块的四个元素的索引。blockStart + 3 要保留，我们通过 remove(...) 方法删除所有其他行。对这个方法要小心一点，因为它是超载的。

- 变体 remove(Object) 从列表中删除了一个和 equals(...) 相等的元素。
- 第二个 remove(int) 方法在给定的位置删除了一个条目。

删除元素时，我们从较高的索引向较低的索引移动。对于列表（尤其是 ArrayList），从后面开始删除总是明智的，这样需要在内存中移动的元素更少。如果我们从 blockStart + 0 开始删除，则索引下的元素被删除，所有其他元素都向上移动。解决方案如下所示：

```
lines.remove( blockStart );
lines.remove( blockStart );
lines.remove( blockStart );
```

任务 3.2.3：缩短列表，因为衰退不存在

```
static void trimNonGrowingNumbers( List<Double> numbers ) {

  if ( numbers.size() < 2 )
    return;

  double previous = numbers.get( 0 );
  for ( int i = 1; i < numbers.size(); i++ ) {
    double current = numbers.get( i );
    if ( current <= previous ) {
      numbers.subList( i, numbers.size() ).clear();
```

```
      break;
    }
    previous = current;
  }
}
```

列表 3.8 com/tutego/exercise/util/TrimNonGrowingList.java

乍一看，任务很简单：遍历列表，看下一个元素是否更大。如果不是这样，我们就中断。其特殊之处在于：我们记不得新列表中的元素，但必须修改传递给我们的列表。这意味着，从元素变小的那一点开始，我们必须删除传递的列表直到最后。

但是，List 接口中没有这样的方法。因此，我们必须手动重新编程。以下两种解决方案是可能的：

- 编写一个 remove(int) 方法，我们可以将索引传递给该方法，以便可以在该点删除元素。明智的做法是从右向左开始调用该方法，直到我们修剪完列表。
- 利用了数据结构的一个特点，即存在的实时视图。建议解决方案也采用这种方法。subList(...) 方法返回列表的视图，但这是实时的，对此子列表的更改将应用于原始列表，即写入。如果我们用 clear( ) 删除这个子列表，则原始列表中的所有元素也会消失。

任务 3.2.4：和朋友一起吃饭：比较元素，找到共同点

```
public static class Guest {
  public boolean likesToShoot;
  public boolean likesToGamble;
  public boolean likesBlackmail;

  public Guest( boolean likesToShoot, boolean likesToGamble,
      boolean likesBlackmail ) {
    this.likesToShoot = likesToShoot;
    this.likesToGamble = likesToGamble;
    this.likesBlackmail = likesBlackmail;
  }

  public boolean hasDissimilarInterests( Guest other ) {
    return !(likesToShoot   == other.likesToShoot ||
```

```java
                    likesToGamble  == other.likesToGamble ||
                    likesBlackmail == other.likesBlackmail);
  }
}

public static int allGuestsHaveSimilarInterests( List<Guest> guests ) {
  for ( int index = 0; index < guests.size(); index++ ) {
    Guest guest = guests.get( index );
    Guest rightNeighbor = guests.get( (index + 1)% guests.size() );
    if ( guest.hasDissimilarInterests( rightNeighbor ) )
      return index;
  }
  return -1;
}
```

列表 3.9 com/tutego/exercise/util/FriendsSittingTogether.java

我们想要将客人表示为具有三个属性的对象。这是由 Guest 类完成的，除了参数化的构造函数外，它还有一个 hasDissimilarInterests(Guest) 方法；它将自己与另一个 Guest 进行比较，并检查是否有共同兴趣。

使用 allGuestsHaveSimilarInterests(...) 方法，一个循环会遍历所有客人。同时我们查询客人及其右侧邻居，列表中的最后一个元素后面没有元素。使用余数运算符回到列表的前面，即将最后一个元素与第一个元素进行比较。没错，因为所有客人都围坐成一圈，每个客人都有一个邻居。

在循环中，hasDissimilarInterests(...) 确定两个客人是否有相同的兴趣。在这种情况下，从我们的方法返回索引。如果循环遍历所有客人并且他们都有共同点，则返回 –1。

任务 3.2.5：检查列表的相同元素顺序

让我们看看两种建议解决方案。两者都根据相同的基本原理工作，即原始列表加倍，当查询在最后一个元素之后结束时，它会从第一个元素重新开始。

如果我们将任务列表加倍，则会看到第二个列表出现在其中：

"Alexandre", "Charles", "Anne", "Henry", "Alexandre", "Charles", "Anne", "Henry"

建议解决方案 1（复制列表）如下：

```java
    List<String> names1 = Arrays.asList( "Alexandre", "Charles", "Anne",
"Henry" );
    List<String> names2 = Arrays.asList( "Anne", "Henry", "Alexandre",
"Charles" );

    ArrayList<String> duplicatedList = new ArrayList<>( names1 );
    duplicatedList.addAll( names1 );
    System.out.println( Collections.indexOfSubList( duplicatedList,
names2 ) >= 0 );
```

列表 3.10 com/tutego/exercise/util/SameInTheCircle.java

对于建议解决方案 1，将所有名字复制到一个新列表中，因为我们不想破坏原始列表，并且该列表甚至可能是不可变的。使用 addAll(...) 方法，我们再次将源中的相同元素应用于此副本。这样，我们将第一个列表的内容翻倍。表达式 Collections.indexOfSubList(List<?> source, List<?> target) 实现测试并返回列表 target 在列表 source 中出现的位置。我们从方法中得到的不是逻辑值，而是直接得到位置，只需要检查这个位置是否大于等于 0。如果第一个列表中不存在第二个列表，则该方法返回 −1。

我们不能使用 containsAll(...)，因为该方法不检查顺序，而只检查第二个集合的所有元素是否出现在第一个集合中，它完全独立于顺序，但顺序至关重要。

建议解决方案 2（虚拟重复列表）如下。

我们需要的是一个大小为第一个列表 2 倍的列表，并且在访问最后一个元素之后从第一个元素重新开始。

```java
  private static <T> boolean isSameCircle( List<T> list1, List<T> list2 ) {

    if ( list1.size() != list2.size() )
      return false;

    AbstractList<Object> list1Duplicated = new AbstractList<>() {
      @Override public int size() {
        return list1.size() * 2;
      }

      @Override public Object get( int index ) {
        return list1.get( index% list1.size() );
```

```
    }
  };
  return Collections.indexOfSubList( list1Duplicated, list2 ) >= 0;
}
```

列表 3.11 com/tutego/exercise/util/SameInTheCircle.java

该方法首先检查列表的大小是否相同。此查询在建议解决方案 1 中也是必需的。由于 isSameCircle(...) 原则上可以处理所有对象而不仅是字符串，所以该方法被声明为静态泛型方法。进行比较时，只需要一个有效的 equals( ) 方法实现即可。

虚拟列表由 AbstractList 的子类实现。这个基类通常用于列表实现，以便尽可能多地采用标准功能。我们重写了两个方法：

- size( ) 方法返回的元素数量是原始列表的 2 倍。
- get(int) 方法通过余数运算符实现，如果我们超出索引的大小，则将从数据结构的开头重新开始。

建议解决方案 2 实际上并没有在内存中创建一个新的列表，它是虚拟的，对外人来说是双倍的。修改方法根本不起作用，但是 size( ) 和 get(...) 再次与 indexOfSubList(...) 一起工作。

任务 3.2.6：现在播报天气：寻找重复元素

```
public static class WeatherOccurrence {
  public String weather;
  public int occurrences;
  public int startIndex;

  WeatherOccurrence( String weather, int occurrences, int startIndex ) {
    this.weather = weather;
    this.occurrences = occurrences;
    this.startIndex = startIndex;
  }

  @Override public String toString() {
    return "weather='" + weather + "', " +
           "occurrences=" + occurrences + ", startIndex=" + startIndex;
```

```java
    }
  }

  static WeatherOccurrence longestSequenceOfSameWeather( List<String> weather ) {

    int localMaxOccurrences = 1;
    int localStartIndex     = 0;

    int globalMaxOccurrences = localMaxOccurrences;
    int globalStartIndex     = localStartIndex;

    String recurringElement = weather.get( 0 );

    for ( int i = 1; i < weather.size(); i++ ) {
      String currentElement = weather.get( i );

      if ( Objects.equals( currentElement, recurringElement ) ) {
        localMaxOccurrences++;
        if ( localMaxOccurrences > globalMaxOccurrences ) {
          globalMaxOccurrences = localMaxOccurrences;
          globalStartIndex     = localStartIndex;
        }
      }

      else { // currentElement != recurringElement
        localStartIndex = i;
        localMaxOccurrences = 1;
        recurringElement = currentElement;
      }
    }

    return new WeatherOccurrence(
        weather.get( globalStartIndex ), globalMaxOccurrences, globalStartIndex );
  }
```

列表 3.12 com/tutego/exercise/util/WeatherOccurrences.java

任务中声明的 WeatherOccurrence 类增加了两个东西：一个用于快速设置状态的参数化构造函数，以及一个 toString( ) 方法。

为了完成这个任务，必须考虑两个不同的序列：一个局部最长的序列和一个全局最长的序列。因此，我们需要一系列可以记住状态的变量。两个变量存储相同元素的局部最大数量和该序列的起始索引，另外两个变量存储找到的全局最大元素数量及其位置。

程序必须回答一个元素是否连续出现多次的问题。我们在开头用第一个元素初始化变量 recurringElement，然后看这个元素是否重复。实际循环可以从索引 1 开始。循环体读取元素并将其与 recurringElement 进行比较，以查看是否存在 recurringElement 序列。Object 类的静态方法 equals(...) 执行相等性测试，因为这比在元素上调用 equals(...) 有优势，即 null 不会导致问题。如果该元素重复出现，则计数器 localMaxOccurrences 会增加 1，并且条件判断检查本地同样出现的元素数量是否超过了全局最大值。如果出现这种情况，那么 if 块就会更新变量 globalMaxOccurrences 和 globalStartIndex。不必记住实际元素本身，因为这只在最后有用，我们可以查询该元素，因为我们知道该元素在数据结构中的位置。

如果列表中出现非等值对象，则将新序列的起始位置放在索引上，将重复元素的数量设置为 1，并重新初始化 recurringElement 以作进一步处理。

在循环结束时，参数化构造函数创建具有 3 个所需状态的 WeatherOccurrence。

任务 3.2.7：创建收据输出

```java
public class Receipt {
  public static class Item {
    public final String name;
    public final int centPrice;
    public final int occurrence;

    public Item( String name, int centPrice, int occurrence ) {
      if ( centPrice <= 0 ) throw new IllegalArgumentException(
        "Price can not be <= 0" );
      if ( occurrence <= 0 ) throw new IllegalArgumentException(
        "Occurrence can not be <= 0" );
      this.name = Objects.requireNonNull( name );
      this.centPrice = centPrice;
      this.occurrence = occurrence;
    }
```

```java
        public Item( String name, int centPrice ) {
          this( name, centPrice, 1 );
        }

        public Item incrementOccurrence() {
          return new Item( name, centPrice, occurrence + 1 );
        }

        @Override public boolean equals( Object other ) {
          if ( other == null || getClass() != other.getClass() )
            return false;

          return centPrice == ((Item) other).centPrice
            && name.equals( ((Item) other).name );
        }

         @Override public int hashCode() {
           return name.hashCode() * 31 + centPrice;
         }
    }

    private final List<Item> items = new ArrayList<>();

    public void addItem( Item item ) {
       int maybeIndex = items.indexOf( item );

        if ( maybeIndex >=0 )
           items.set( maybeIndex, items.get( maybeIndex ).incrementOccurrence() );
         else
           items.add( item );
    }

    @Override public String toString() {
       NumberFormat currencyFormatter = NumberFormat.getCurrencyInstance(
         Locale.GERMANY );
```

```java
StringBuilder result = new StringBuilder( 512 );
int sum = 0;

for ( Item item : items ) {
  int itemPriceTotal = item.centPrice * item.occurrence;
  String line = String.format( "%dx%-20s%10s%10s%n",
    item.occurrence, item.name,
    currencyFormatter.format( item.centPrice / 100. ),
    currencyFormatter.format( itemPriceTotal / 100. ) );
  result.append( line );
  sum += itemPriceTotal;
}

result.append( "\nSum: " )
      .append( currencyFormatter.format( sum / 100. ) )
      .append( "\n" );

return result.toString();
  }
}
```

列表 3.13 com/tutego/exercise/util/Receipt.java

首先，让我们看一下 Item 类。有两个参数化构造函数：

- ▶ 第一个构造函数初始化所有三个信息——名称、价格和出现次数。构造函数还检查有效性。
- ▶ 第二个构造函数只是数字 1 的简化。

由于 Item 对象是不可变的，所以 incrementOccurrence( ) 方法会返回一个增加了数量的新 Item 对象。Item 类覆盖 Object 的两个方法：equals(...) 和 hashCode( )。比较方法稍后会很重要，因为我们可以使用它来搜索数据结构中的 Item 对象。没有考虑 occurrence，我们马上就会知道为什么。

收据本身由 Item 对象的集合组成。使用 addItem(Item) 方法，我们向收据添加一个新 Item。我们不只是要追加条目，如果已经存在具有相同名称和数量的条目，那么我们希望进行缩减，然后将其合并。首先，indexOf(...) 从列表中搜索和 equals(...) 相等的 Item，因此必须忽略 occurrence。如果 indexOf(...) 找到一个元素，

则结果 >= 0，即找到的元素的索引。其后的条件判断区分如下：

- 如果找到了该元素，则该元素将在其位置被新元素替换，occurrence 增加 1。
- 如果在列表中没有找到与 equals(...) 相等的 Item，则将其附加在后面。

最后，来到 toString() 方法。它必须遍历所有条目并输出产品名称、数量、价格和数量与价格的乘积。我们可以通过不同的方式实现价格输出，此处选择的解决方案使用 NumberFormat 类，以便能够长期支持不同语言的货币符号。用 NumberFormat.getCurrencyInstance(Locale.GERMANY) 创建的对象会自动将欧元符号放在数字后面。计算总和也是 toString() 方法的任务，然后可以在最后输出。除以 100 时，要确保不是整数除法，否则会漏掉小数位。配置的 NumberFormat 自动设置两位小数并用 0 填充。

测试 3.2.8：装饰数组
结果如下：

```
Exception in thread "main" java.lang.UnsupportedOperationException
```

asList(...) 实现了适配器设计模式，它使两个不兼容的 API 相互适应。在这种情况下，它将带大括号的类型数组改编为用于读写，并将属性 length 改编为 java.util.List。列表上的操作是实时的，并被写入数组。没有列表被创建为副本。由于数组的长度在创建后不能改变，所以元素既不能被删除也不能被添加。如果试图这样做，就会出现 UnsupportedOperationException。这在最初的实现中可以看得很清楚：

```
public static <T> List<T> asList( T... a ) {
  return new ArrayList<>( a );
}

private static class ArrayList<E> extends AbstractList<E>
    implements RandomAccess, java.io.Serializable {

  private final E[] a;

  ArrayList( E[] array ) {
    a = Objects.requireNonNull( array );
```

```
    }

    @Override
    public int size() { return a.length; }

    @Override
    public E get( int index ) { return a[ index ]; }

    ...
}
```

列表 3.14 Aus der OpenJDK-Implementierung von java.util.Arrays

```
public void add( int index, E element ) {
    throw new UnsupportedOperationException();
}
```

列表 3.15 Aus der OpenJDK-Implementierung von java.util.AbstractList

asList(T...a) 方法创建 ArrayList 类型的实例，这并不是 java.util.ArrayList，而是 Arrays 中的嵌套类。很容易看出，阅读方法直接进入数组，但是 add(...) 方法抛出 UnsupportedOperationException。

测试 3.2.9：查找和未找到
输出如下：

```
false
true
true
```

这里使用的方法声明如下：

```
public static <T> List<T> asList(T... a)
```

参数类型是可变参数，即任何对象数组。在第一种情况下，我们将原始整数数组传递给 asList(...)，原始数据类型不是由泛型类型变量处理的引用类型。这意味着类型 T 代表引用类型 "int[];"，因此，产生的列表类型为 List<int[]>，并且 contains(...) 方法返回 false。

JVM 内部没有可变参数，它们是普通的数组，不同的是我们可以在调用时直接传递一些参数，我们将它们收集并传递到一个内部数组中。因此，有两种方法可以使用可变参数方法：枚举多个参数，然后将其打包到一个内部数组中，或者直接传递一个数组。这正是我们在这里使用的变体。包装类型 Integer 是类型变量 T 的类型参数；创建了一个 List<Integer> 并且 contains(...) 找到 1，因为 1 通过装箱变成 Integer 对象，并且包装器对象实现 equals(...)。

在第三部分，我们使用可变参数列表。首先将 int 元素装箱成整数对象，然后放入匿名内部数组并传递。

任务 3.2.10：加上奶酪，一切都会更美味：在列表中添加元素

谈到数据结构的变化，我们基本上可以走两条路：创建具有所需属性的新数据结构或自己修改数据结构。所有典型的数据结构，如 ArrayList、HashMap、TreeSet 都可以改变，这就是这个任务的方法，将元素添加到现有的数据结构中。

该任务讲的是一个列表，列表中的元素有一个位置。如果必须在某个位置引入某些东西，基本上有两种可能：可以使用面向索引的方法 add(int index, Eelement) 或通过 ListIterator 插入元素。面向索引的方法在 LinkedList 中有一个缺点，因为它不能通过索引提供对元素的快速 RandomAccess。因为该方法具有通用参数类型 List，所以我们必须调整到每个可能的列表，并且不希望 LinkedList 的性能不佳。就输出而言，ListIterator 是完美的。

```java
public static void insertCheeseAroundVegetable( List<String> ingredients ) {
  final String VEGETABLE = "Zucchini|Paprikas?|Zwiebeln?|Tomaten?";
  for ( ListIterator<String> iterator = ingredients.listIterator();
    iterator.hasNext(); ) {
    String ingredient = iterator.next();
    Pattern p = Pattern.compile( VEGETABLE );
    Matcher m = p.matcher( ingredient );
    if ( m.matches() )
      // The new element is inserted before the implicit cursor
      iterator.add( "Käse" );
  }
}
```

列表 3.16 com/tutego/exercise/util/CheeseInserter.java

建议的解决方案从列表中获取 ListIterator，并像往常一样使用 hasNext() 和 next() 的组合遍历所有元素。提取元素后，matches(...) 会询问该元素是否与预定义

的字符串之一匹配。因此，对于这个检查，我们使用一个包含不同类型蔬菜的正则表达式，甚至在选定品种上使用复数 s 也是正确的。如果字符串匹配，则插入"Käse"。

我们可以将迭代器视为位于元素之间的游标。在 next( ) 之后，光标位于元素后面。add(...) 方法将元素插入光标前面，即在 next( ) 之后返回的元素后面。换句话说，光标不是被放在新插入元素的前面，而是被放在它后面。也就是说，下一个 next( ) 不会返回新内置的"Käse"，而是返回下一个常规成分。

测试 3.2.11：一无所有让人很恼火

简而言之：不。有一个异常：

```
Exception in thread "main" java.util.ConcurrentModificationException
    at java.base/
java.util.ArrayList$Itr.checkForComodification(ArrayList.java:1012)
    at java.base/java.util.ArrayList$Itr.next(ArrayList.java:966)
```

程序看起来很无辜：扩展的 for 循环遍历列表，我们删除列表中的一个元素。乍一看，删除元素会导致 ConcurrentModificationException，这很奇怪。concurrent 在这里应该是什么意思？我们没使用线程！

在我们的示例中，concurrent（中文：同时）意味着迭代数据结构并"同时"删除它。当迭代器继续运行时，Java 会识别出数据结构已经发生了变化。换句话说：迭代和删除由状态耦合。它的工作原理是这样的：扩展的 for 循环只在内部使用迭代器。如果我们改写程序，就像真的在字节码中，那么我们会得到以下内容：

```
for ( Iterator<String> iterator = names.iterator(); iterator.hasNext(); ) {
  String name = iterator.next();
  if ( "".equals( name ) )
    names.remove( name );

  System.out.println( names );
  }
}
```

List 和 Iterator 的实现有一个状态，它们会记住修改的数量。创建 Iterator 时，它会使用列表的修改计数对其自身进行一次初始化，并且随着列表的更改，它会增加其修改计数。稍后，当 Iterator 移动到下一个元素时，它将其存储的修改计数与列表的修改计数进行比较，如果不匹配则抛出异常。在代码中它看起来像这样：

```
class AbstractList {
  protected transient int modCount = 0;

  public Iterator<E> iterator() {
    return new Itr();
  }

  private class Itr implements Iterator<E> {
    int expectedModCount = modCount;

    public E next() {
      checkForComodification();
      ...
    }

    final void checkForComodification() {
      if (modCount != expectedModCount)
        throw new ConcurrentModificationException();
    }
    ...
  }
  ...
}
```

结果是：我们无法对迭代循环内的数据结构进行任何更改。运行和删除不能那样做，此类任务需要其他解决方案。没有循环的解决方案可能如下所示：

```
names.removeIf( ""::equals );
```

removeIf(...) 需要一个谓词，而这个漂亮又紧凑的方法引用实现了这样一个谓词，用于测试下一对象是否与空字符串 equals(...) 相等。

任务 3.2.12：使用迭代器搜索元素，找到 Covid Cough

我们得到一个初始化的 Iterator，它只能左、右运行，不能绝对跳跃——就像人一样受到限制。选择正确的策略并不容易，因为我们可以在这里通过概率或尝试降低最大"成本"。

策略 1：假设我们有 100 艘船，而 Bonny Brain 站在第 90 艘船上。这意味着一个方向有 90 艘船，你要找的人有 90% 的机会在那里。但是，有 10% 的可能性是你要找的人不在该区域，然后 Bonny Brain 必须一直向下滚动列表才能查看其余部分。在最坏的情况下，必须访问 90 + 90 + 10 = 190 艘船。

策略 2：Bonny Brain 知道左侧或右侧是否有更多船。该策略是先走船少的一边，目的是给最坏的情况降低总体成本。对于有 100 艘船的例子，Bonny Brain 站在第 90 艘船上，这意味着 10 + 10 + 90 = 110 次访问（最多）。

建议解决方案实现了策略 2：

```
if ( iterator.nextIndex() >= NUMBER_OF_SHIPS / 2 ) {
  if ( searchRight( iterator ) )
    System.out.println( "-> at ship " + iterator.previousIndex() );
  else if ( searchLeft( iterator ) )
    System.out.println( "-> <- at ship " + iterator.nextIndex() );
  else
    System.out.println( "Not found" );
}
else {
  if ( searchLeft( iterator ) )
    System.out.println( "<- at ship " + iterator.nextIndex() );
  else if ( searchRight( iterator ) )
    System.out.println( "<- -> at ship " + iterator.previousIndex() );
  else
    System.out.println( "Not found" );
}
```

列表 3.17 com/tutego/exercise/util/FindCovidCough.java

Iterator 方法 nextIndex( ) 返回绝对位置，我们用它来判断 Bonny Brain 是在右半边还是左半边。自己的方法 searchRight(...) 首先使用 Iterator 运行右侧，并在找到搜索对象时以 true 响应。如果方法返回 false，则表明我们在右半边没有找到搜索对象，我们必须从那里再次往回运行，一直运行到左边。输出中的箭头指示搜索方向。如果我们到达左侧并且 searchLeft(...) 返回 false，那么搜索对象根本不在列表中！第二个大 else 分支测试另一种情况，CiaoCiao 船长相对于中心站在左侧，Bonny Brain 首先一直朝左边走。

searchRight(Iterator<Ship>) 和 searchLeft(ListIterator<Ship>) 是这样实现的：

```java
final static String COVID_COUGH = "Covid Cough";

private static boolean searchRight( Iterator<Ship> iterator ) {
  while ( iterator.hasNext() )
    if ( iterator.next().contains( COVID_COUGH ) )
      return true;
  return false;
}

private static boolean searchLeft( ListIterator<Ship> iterator ) {
  while ( iterator.hasPrevious() )
    if ( iterator.previous().contains( COVID_COUGH ) )
      return true;
  return false;
}
```

列表 3.18 com/tutego/exercise/util/FindCovidCough.java

向右寻找的方法不需要 ListIterator，因为普通的 Iterator 提供了两个方法，hasNext( ) 和 next( )，向右移动并提取下一个元素。searchLeft(...) 方法需要一个 ListIterator，因为我们需要使用 hasPrevious( ) 和 previous( ) 方法向左运行。一个普通的 Iterator 不能任意向左或向右运行。只有使用 ListIterator，我们才能多次遍历数据结构。

任务 3.2.13：移动元素，玩"抢椅子"游戏

```java
class MusicalChairs {

  private final List<String> names;

  public MusicalChairs( String... names ) {
    if ( names.length == 0 )
      throw new IllegalArgumentException(
          "no names are given, but names must not be empty" );
    this.names = new ArrayList<>( Arrays.asList( names ) );
  }
```

```java
  public void rotate( int distance ) {
    Collections.rotate( names, distance );
  }

  public void rotateAndRemoveLast( int distance ) {
    if ( names.isEmpty() )
      throw new IllegalStateException(
        "names is empty, no names to remove" );

    rotate( distance );
    names.remove( names.size() - 1 );
  }

  public String play() {
    if ( names.isEmpty() )
      throw new IllegalStateException(
        "names is empty, no names to play with" );

    while ( names.size() > 1 ) {
      rotateAndRemoveLast( ThreadLocalRandom.current().nextInt() );
      System.out.println( names );
    }

    return names.get( 0 );
  }

  @Override public String toString() {
    return String.join( ", ", names );
  }
}
```

列表 3.19 com/tutego/exercise/util/MusicalChairsGame.java

MusicalChairs 类有一个 List 类型的对象变量，其中包含名称。尽管在构造函数中传递了一个可变参数数组，但如果我们以后想使用 Collections 类的 rotate(...) 方法，列表也给我们提供了更高的灵活性。构造函数将数组转换为列表，并预先检查数组是否包含任何元素——否则构造函数会抛出异常。如果使用 null 调用构造函数，则会抛出通常的 NullPointerException。

三种办法：

- rotate(...) 方法利用了 Collections 的 rotate(...) 方法。这个方法有效，它修改了列表。我们的名单是内部的，不能从外部进入。
- rotateAndRemoveLast(...) 方法首先执行旋转，然后删除最后一个列表元素，但是有可能报错：如果游戏进行了多轮，则重复调用 rotateAndRemoveLast(...) 可以使列表为空。第一种条件判断检查这种情况并在列表为空时抛出异常。如果有多个元素，我们会旋转列表并从列表中删除最后一个项目。ArrayList 没有删除最后一个元素的专用方法。如果我们使用 LinkedList 而不是 ArrayList，那么 Deque 接口就会有一个 removeLast( ) 方法。
- play( ) 方法运行游戏，同样，列表不能为空。while 循环执行主体，直到列表中的元素数变为 1。如果列表中有多个名称，则轮换列表并输出。循环运行后，列表只包含一个元素。我们回到第一个元素。同样，List 没有提供获取第一个元素的特定方法。实现 Queue 接口的数据结构则不一样：这里有 remove( ) 方法。
- toString( ) 方法使用 String.join(...) 静态方法，因为这是创建包含元素的以逗号分隔的字符串的最简单的方法。

任务 3.2.14：编辑行星的问答游戏

```
List<Planet> shuffledPlanets = new ArrayList<>( Arrays.asList(
  Planet.values() ) );
Collections.shuffle( shuffledPlanets );

for ( Planet question : shuffledPlanets ) {
  System.out.printf( "What is the diameter of planet%s (in km)?%n",
    question.name );

  List<Planet> misleadingPlanets =
    new ArrayList<>( Arrays.asList( Planet.values() ) );
  misleadingPlanets.remove( question );
  Collections.shuffle( misleadingPlanets );

  List<Planet> choicePlanets = misleadingPlanets.subList( 0, 3 );
  choicePlanets.add( question );
```

```java
    Collections.shuffle( choicePlanets );
    choicePlanets.forEach( planet ->
      System.out.println( planet.diameter + " km" ) );

    if ( new Scanner( System.in ).nextInt() != question.diameter )
      System.out.printf( "Wrong! The diameter of%s is%d km.%n%n",
         question.name, question.diameter );
    else
      System.out.printf( "Correct!%n%n" );
}
```

列表 3.20 com/tutego/exercise/util/PlanetQuiz.java

建议解决方案分不同阶段进行。在第一阶段，创建一个新 List，然后使用 shuffle(...) 方法进行混洗。现在，通过扩展的 for 循环，我们可以遍历这些杂乱无章的行星，针对每个行星提出一个问题，该问题必须被回答。

下一步，从已知行星中随机选择三颗行星。请注意，被提问过的行星不会再次出现在随机选择中。因此，我们另建列表，删除问题的答案［枚举元素实现 equals（Object）方法适用于此］，打乱列表，从列表中选择三个行星，将答案附加到列表中并再次打乱列表。这保证了答案不会一直出现在同一个位置，并且三个备选答案是不同的。

屏幕上的输出和问题不会消失。如果行星直径不正确，就会显示正确答案。

任务 3.3.1：形成子集，寻找共同点

该任务寻找的交集不能直接通过单一的运算生成。Collections 或 Set 以及集合返回的实现中都没有方法，该集合包含两个集合的元素，因此我们绕道而行。

```java
Set<String> me = new HashSet<>();
Collections.addAll( me, "Candy making", "Camping", "Billiards",
   "Fishkeeping", "Eating", "Action figures", "Birdwatching", "Axe throwing" );
Set<String> she = new HashSet<>();
Collections.addAll( she, "Axe throwing", "Candy making", "Camping",
   "Action figures", "Casemodding", "Skiing", "Satellite watching" );

Set<String> smallerSet, largerSet;
if ( me.size() < she.size() ) {
  smallerSet = me; largerSet = she;
```

```
    } else {
      smallerSet = she; largerSet = me;
    }

    Set<String> intersection = new HashSet<>( smallerSet );
    intersection.retainAll( largerSet );
    System.out.println( intersection );

    System.out.printf( "Liste 'me' stimmt zu%d%% mit der Liste 'she'
überein.%n", (intersection.size() * 100) / me.size() );
    System.out.printf( "Liste 'she' stimmt zu%d%% mit der Liste 'me'
überein.%n", (intersection.size() * 100) / she.size() );
```

列表 3.21 com/tutego/exercise/util/DatingCompatibility.java

第一步，创建两个集合之一的副本。就其工作方式而言，复制哪个集合并不重要；我们选择较小的那个集合，这个原因稍后探讨。复制的原因是接下来的方法修改了集合，而我们在任何情况下都不想修改传入的集合；可能传入方法的集合也是不可变的，这意味着会发生异常。

创建副本后，我们使用 Set 方法 boolean retainAll(Collection<?>)。它修改方法调用的对象，只留下既存在于自己的集合中，也存在于传递集合中的元素。这实际上是一个交集的构成。签名有两个有趣的地方：

1. 参数不是一个集合，而是一个任意 Collection。也就是说，我们也可以传递一个列表。即使列表中有元素多次出现，也无关紧要。
2. 另外，retainAll(Collection<?>) 的 <?> 很有趣，这说明转移集合的类型无关紧要。在内部，对象与 equals(...) 进行比较。

retainAll(...) 的实现很简单：

```
public boolean retainAll( Collection<?> c ) {
  Objects.requireNonNull( c );
  boolean modified = false;
  Iterator<E> it = iterator();
  while ( it.hasNext() ) {
    if ( !c.contains( it.next() ) ) {
      it.remove();
```

```
            modified = true;
        }
    }
    return modified;
}
```

列表 3.22 Aus der OpenJDK-Implementierung von java.util.AbstractCollection

标准实现使用 Iterator 遍历自己的集合，并使用 contains(...) 询问元素是否出现在传递的集合中。如果出现，则通过 Iterator 删除该元素。如果传递的 Collection 是一个列表，那么 contains(...) 方法平均比在 TreeSet 或 HashSet 上查询更昂贵。

知道了这一点，我们回到刚才的问题，我们复制了两个集合中较小的那一个。这样做有两个后果：

1. 一方面复制小的数据结构比复制大的集合更快、更节省内存。
2. 另一方面，retainAll(...) 的 OpenJDK 实现告诉我们，迭代器在自己的集合上运行，这意味着如果它自己的集合比较小，那么访问的元素就比较少。我们可以反驳说，总是有必要对更大的（或同样大的）集合进行测试，但如果总体上查询次数较少，那么平均来说就会更快。

形成交集后，我们把交集的大小与两个人的爱好数量联系起来。如果这两个人有不同数量的爱好，则匹配的百分比是不一样的；只有在双方都表示有相同数量的爱好时，匹配的百分比才相同。在我们的例子中，程序输出如下：

```
[Candy making, Axe throwing, Camping, Action figures]
Liste 'me' stimmt zu 50% mit der Liste 'she' überein.
Liste 'she' stimmt zu 57% mit der Liste 'me' überein.
```
（列表 me 和列表 she 的匹配度为 50%。
列表 she 和列表 me 的匹配度为 57%。）

测试 3.3.2：好剑
输出如下：

```
true false [Khanda]
```

没有异常。

Java 中有两种排序方式：要么是对象相互比较，然后类实现 Comparable 接口；要么是 Comparator 查看两个对象并决定两个对象的顺序。在我们的示例中，该类以自然顺序实现了 Comparable 接口，但 compareTo(...) 方法始终返回 0。这意味着所有对象都将是相同的，无论它们的状态 name 是什么。分配的名称根本不在比较方法中。

将元素插入 Set 时，仅当集合中尚无等价元素时才接受该元素。数据结构通常使用 equals(...) 方法。TreeSet 是一个有序集合，是一个异常，因为实现不需要 equals(...) 方法，可以使用 compareTo(...) 读取两个对象是否相等。是否实现单独的 equals(...) 方法或是否引发异常都没有关系，因为在该场景中不会调用 equals(...)。

main(...) 方法输出三样东西：两次添加方法 add(...) 的返回值和通过 toString() 表示的集合内容。集合的 add(...) 方法返回一个 boolean 值，表示是否将元素收入集合。在第一种情况下，添加了一个字符串，因此返回 true。第二次调用 add(...) 方法时，TreeSet 进行识别并根据 Comparable 的第二个字符串使其等价于集合中的现有元素。相同的元素不会被覆盖和替换，但该元素会被丢弃。由于没有添加任何内容，所以 add(...) 方法返回 false。在集合的 toString() 表示中，仅出现第一个插入的元素。

任务 3.3.3：删除数组中的重复元素

```java
public static double[] unique( double... values ) {

  if ( values.length < 2 )
    return values;

  Set<Double> valuesSet = new LinkedHashSet<>( values.length / 4 );

  for ( double value : values )
    valuesSet.add( value );

  if ( valuesSet.size() == values.length )
    return values;

  double[] result = new double[ valuesSet.size() ];
  int i = 0;
  for ( Double value : valuesSet )
    result[ i++ ] = value;
```

```
    return result;
}
```
列表 3.23 com/tutego/exercise/util/UniqueArrayElements.java

使用 LinkedHashSet 数据结构可以轻松实现该解决方案。LinkedHashSet 的优点是一方面保留了插入元素的顺序，另一方面数据结构仍然是一个集合，并且只包含元素一次。由于进行了一些特殊检查，所以解决方案要长几行，但这些优化并没有错。

第一步，该方法检查数组中是否没有元素或只有一个元素。如果是这样，那么我们直接返回数组。此外，通过访问属性 length，我们可以检查 null，因为如果传递 null 会引发 NullPointerException。

如果有两个以上的元素，则可能有重复。我们现在才需要创建 LinkedHashSet 的实例。估计大小很困难，因为我们不知道是否有许多元素重复出现，因此，初始大小的四分之一可能太多或太少，实际测试必须证明这一点。

下一步，扩展的 for 循环遍历原始数组并将每个值放入数据结构，同时装箱，即 double 类型的原始值被包装在 Double 类型的包装对象中。与整数值不同，Double 不对 valueOf(...) 执行任何优化，也不使用缓存。因此，包装对象的数量总是与数组中的元素一样多。

填充 LinkedHashSet 之后，我们的方法将数据结构的大小与数组的大小进行比较，如果两者的元素数量相同，则所有元素都不同。在这种情况下，我们可以直接返回数组，忽略数据结构。

如果数据结构中的元素少于数组中的元素，则必须创建一个包含与集合一样多元素的小数组。这一次，扩展的 for 循环在数据结构上运行，Java 编译器通过使用 Iterator 并调用 hasNext() 和 next() 自动实现。LinkedHashSet 对象的 Iterator 考虑到了插入元素的顺序。由 Iterator 提供的 Double 对象通过拆箱被拆开，并作为 double 条目按顺序被放在数组中。

任务 3.3.4：查明单词中包含的所有单词

```java
private static final int MIN_WORD_LENGTH = 3;

private static Collection<String> substrings( String string ) {
  Collection<String> result = new ArrayList<>(
      (int)(string.length() * (string.length() - 3L) / 2 + 1) );

  for ( int startIndex = 0; startIndex < string.length(); startIndex++ )
    for ( int len = MIN_WORD_LENGTH; len <= string.length() - startIndex;
```

```
      len++ )
    result.add( string.substring( startIndex, startIndex + len ) );

  return result;
}

public static Collection<String> wordList( String string,
  Collection<String> words ) {
  Collection<String> result = new ArrayList<>();

  for ( String substring : substrings( string.toLowerCase() ) )
    if ( words.contains( substring ) )
      result.add( substring );

  return result;
}
```

列表 3.24 com/tutego/exercise/util/WordSequence.java

在实现带有字典访问的 wordList(...) 方法之前,应该实现另一个方法:substrings (String)。该方法返回所有可能子字符串的集合。该方法的核心由两个嵌套循环组成。外循环指定起始位置,内循环生成从至少 3 个字符到最大字符串长度的所有长度。最小尺寸由一个常量决定,因此很容易更改。

结果建立在一个内部列表中。列表的大小可以被计算出来。此处显式类型转换的背景是在将两个大的 int 值相乘时超出了值的范围。用 3L 做减法时,结果会被强制转换为 long 类型,然后由显式类型转换将 long 值更改回 int 值。

使用 String 方法 substring(...) 创建子字符串并将其添加到结果集中。我们选择 ArrayList 作为数据结构,因为它节省了大量空间,之后会从前到后按顺序运行。此外,我们可以提前确定期望的元素数量,这样内部数组就不必在运行时扩大。

有了前期工作,wordList(...) 的实现很短。同样,一个 ArrayList 被建立为容器并返回。在循环中,当一个词在传递的字典中出现时,这个词就被放在这个容器中。

任务 3.3.5:正确分类几乎相同的东西

再次总结一下这项任务的要求:输入是由必须按照一个顺序排列的字符串组成的。顺序是由字符串的数值决定的。然而,即使字符串是不相同的,数字表示也可能是相同的。

对于排序,我们可以求助像 TreeSet 这样的排序数据结构,或者把元素放在一个列表中，然后进行排序。无论我们选择哪种方法，都需要一个特殊的排序标准。对此，Comparator 派上了用场。

```
Comparator<String> comparator =
    Comparator.comparing( (String s) -> new BigDecimal(s))
              .thenComparing( Comparator.naturalOrder() );

SortedSet<String> sortedNumbers = new TreeSet<>( comparator );
Collections.addAll( sortedNumbers,
    "-13.123", "0", "0", "10101010", "10101010.0", "0.0", "-0.0" );

System.out.println( sortedNumbers );
```
列表 3.25 com/tutego/exercise/util/SortEquallyBigNumbers.java

Comparator 需要考虑两件事。如果将字符串转换为数字表示，并且这些值实际上更大或更小，则已经做了决定。但是，如果转换产生等价的结果，那么我们需要引入另一个排序标准，那就是字符串的自然排序。巧合的是，减号在字典顺序中的数字之前，因此 –0.0 和 0 以正确的顺序出现。

可以使用静态方法 comparing(...) 来创建 Comparator。该方法需要一个提取器。在我们的例子中，不需要从复杂对象中提取任何内容，而是将 String 转换为 BigDecimal。结果是一个 Comparator<String>，它可以正确比较数值，但对于两个相等的数值则返回 0。在这种情况下，我们必须解析第二个排序标准。这可以通过 comparing(...) 来完成，因为这里要比较字符串。因为 String 对象具有自然顺序，所以我们可以使用 Comparator.naturalOrder() 来查询使用自然顺序进行字符串比较的 Comparator。

使用 Comparator，我们可以创建一个排序集合。将包含数字的字符串添加到该集合中，并使用 toString() 输出该集合，然后我们就能够读取排序。

任务 3.3.6：用 UniqueIterator 排除重复元素

新迭代器是现有迭代器的装饰器，实际数据来自该迭代器。但是，由于原始迭代器可以返回重复提交的元素，所以我们需要记住元素之前是否出现过。有许多选项非常适合这项任务。像 HashSet 这样的数据结构可以快速回答元素是否在集合中。

基本上，我们必须采取的解决方案是，当在新的迭代器中获得一个元素时，我们必须在后台不断询问内部迭代器，直到返回一个新元素。这本身相对简单，但不能忘记实现 hasNext() 方法。因此，我们需要进行不一样的实现，建议解决方案展示了这一点。

```java
class UniqueIterator<E> implements Iterator<E> {

  private final Iterator<? extends E> iterator;
  private final Set<E> hasSeenSet = new HashSet<>();
  private E next;

  public UniqueIterator( Iterator<? extends E> iterator ) {
    this.iterator = iterator;
    next = lookahead();
  }

  private E lookahead() {
    while ( iterator.hasNext() ) {
      E next = iterator.next();
      if ( ! hasSeenSet.contains( next ) )
        return next;
    }
    return null;
  }

  @Override
  public boolean hasNext() {
    return next != null;
  }

  @Override
  public E next() {
    E result = next;
    hasSeenSet.add( result );
    next = lookahead();
    return result;
  }
}
```

列表 3.26 com/tutego/exercise/util/UniqueIteratorDemo.java

类的构造函数采用原始 Iterator 并将引用存储在内部对象变量中。此外，还有两个对象变量：一个用于已经看到的元素集，另一个引用下一个元素。如果原始 Iterator 不能再返回元素，则 next 将为 null。构造函数还有第二个任务，就是获取第一个元素。这里的重点是 lookahead( ) 方法。

lookahead( ) 方法转到原始 Iterator 并对其进行查询，直到出现一个还没有出现在已经看到的元素集合中的元素。如果 Iterator 不能再返回任何元素，那么新的 UniqueIterator 也不能再返回任何元素，并且该方法返回 null。如果内部 Iterator 找到一个尚未在集合中的元素，则 look ahead( ) 将返回该元素。

总结一下：当构造函数被调用时，立即通过内部 Iterator 查询第一个元素，并将其存储在变量 next 中。如果 next 为 null，则底层 Iterator 中没有元素。

迭代器要覆盖两个方法：

1. hasNext( ) 的实现相应地很简单：如果 next 不等于 null，那么就有一个元素。更新是由 next( ) 完成的。
2. 使用 next( ) 方法，首先标记中间变量 result 中的 next 赋值。如果从外部调用 next( ) 方法，则结果必须被包含在已经看到的元素集中，以便下次认为它是已知的。

对象变量 next 使用 lookahead( ) 方法更新，以便下次调用 Iterator 方法。可能有下一个元素——那么 next 将为 null——或者可能没有元素，则 next 将为 null。更新变量 next 后，next( ) 方法返回的前一个值位于 result 中。

算法就这么多。与许多其他数据类型一样，迭代器是泛型类型。我们也使用这种可能。UniqueIterator 类有一个类型参数 E，它继承自 Iterator 接口。这可以从 next( ) 方法中看出，它返回 E 类型的东西。这个类型变量在构造函数中变得有趣，它接收 Iterator<? extends E>，即不仅期待 Iterator<E>，而且允许更多可能。<? extends E> 表示原始迭代器可以包含 E 类型的子类型。换句话说：如果我们声明一个带有类型参数 Object 的 UniqueIterator，那么内部底层 Iterator 可以返回 String，因为 String extends Object。

任务 3.4.1：将二维数组转换为映射

```
public static Map<String, String> convertToMap( String[][] array ) {

  if ( array.length == 0 )
    return Collections.emptyMap();

  if ( array.length == 1 )
    return Collections.singletonMap( Objects.requireNonNull( array[ 0 ][ 0 ] ),
```

```java
                    Objects.requireNonNull( array[ 0 ][ 0 ] ) );

  Map<String, String> result = new HashMap<>( Math.max( array.length, 16 ) );

  for ( String[] row : array )
    result.put( Objects.requireNonNull( row[ 0 ] ),
                Objects.requireNonNull( row[ 1 ] ) );

  return result;
}
```

列表 3.27 com/tutego/exercise/util/ConvertToMap.java

Java 中的多维数组只不过是引用其他数组的数组。对于二维数组，我们有一个主数组，用于说明列，它为行引用许多小数组。

程序代码有两个特殊之处：

1. 空数组和只有一个键值对的数组的优化；
2. null 校验。

数组是 Java 中的对象，这时候就用到了引用，它们可以为 null。如果将 null 传递给方法，则对主数组长度的第一个查询将抛出 NullPointerException。即使行数组不存在且为 null，通过 array[index] 访问它们也会抛出 NullPointerException。之后对 Objects.requireNonNull(...) 的调用在元素级别测试它们不为 null，否则该方法将引发异常。

关联映射的存储要求比阵列高得多。有两种特殊情况可以显著减少内存占用，一种是不包含任何元素的 Map，另一种是只包含一个键值对的 Map。Collections 类提供了两种特殊的方法来创建空的关联映射和只有一对的关联映射：emptyMap() 和 singletonMap(...)。

只有当元素不只一个时，才建立 HashMap；容量是用数组中的行数预先初始化的，但至少是 16。那么什么是容量？容量是一种缓冲器。一个 HashMap 有一个 DEFAULT_LOAD_FACTOR，默认为 75%；如果添加了新元素并且关联映射达到了 75% 的容量，则进行所谓的重新散列 (Rehashing)。HashMap 被放大，所有元素被重新分类。当然，我们希望降低这些成本。如果我们给出 16 个元素的初始容量，则 HashMap 可以直接容纳 12 个元素而不需要重新散列。知道了这一点，我们也可以用 array.length / 0.75 + 1 来操作，从而计算出最佳的大小。但是，我们假设有一个非常大的数组，那么乘以 1.3 会导致溢出，数字变成负数，然后 HashMap 构造函数中出现异常。这些都是非常极端的特殊情况，但如果我们想成为出色的软件开发者，就必须提防这样的事情。

我们可以使用 HashMap，因为任务提到实现了 equals(...) 和 hashCode( )。如果这两个方法没有得到正确的实现，我们就会有一个问题，数组的所有条目很可能还是会被包含在关联映射中，因为如果 equals(...) 和 hashCode( ) 没有被覆盖，则从超类 Object 来看，每个非相同的对象也不会等同于另一个对象，而且哈希码很可能总是不同的，这样对象之间就毫无关联。

如果数组包含多个元素，则创建 HashMap，运行数组，并将键值对添加到关联映射中。

返回中有一个细微的差别：对于 0 个或 1 个元素，返回是一个不可变的关联映射。如果有 2 个以上元素，那么我们将返回一个可更改的数据结构。

任务 3.4.2：将文本转换为摩尔斯密码并反转

```java
class Morse {
  private final Map<Character, String> charToMorse = new HashMap<>();
  private final Map<String, Character> morseToChar = new HashMap<>();

  Morse() {
    charToMorse.put( 'a', ".-" );
    charToMorse.put( 'b', "-..." );
    charToMorse.put( 'c', "-.-." );
    charToMorse.put( 'd', "-.." );
    charToMorse.put( 'e', "." );
    charToMorse.put( 'f', "..-." );
    charToMorse.put( 'g', "--." );
    charToMorse.put( 'h', "...." );
    charToMorse.put( 'i', ".." );
    charToMorse.put( 'j', ".---" );
    charToMorse.put( 'k', "-.-" );
    charToMorse.put( 'l', ".-.." );
    charToMorse.put( 'm', "--" );
    charToMorse.put( 'n', "-." );
    charToMorse.put( 'o', "---" );
    charToMorse.put( 'p', ".--." );
    charToMorse.put( 'q', "--.-" );
    charToMorse.put( 'r', ".-." );
```

```java
    charToMorse.put( 's', "..." );
    charToMorse.put( 't', "-" );
    charToMorse.put( 'u', "..-" );
    charToMorse.put( 'v', "...-" );
    charToMorse.put( 'w', ".--" );
    charToMorse.put( 'x', "-..-" );
    charToMorse.put( 'y', "-.--" );
    charToMorse.put( 'z', "--.." );
    charToMorse.put( '1', ".----" );
    charToMorse.put( '2', "..---" );
    charToMorse.put( '3', "...--" );
    charToMorse.put( '4', "....-" );
    charToMorse.put( '5', "....." );
    charToMorse.put( '6', "-...." );
    charToMorse.put( '7', "--..." );
    charToMorse.put( '8', "---.." );
    charToMorse.put( '9', "----." );
    charToMorse.put( '0', "-----" );

    charToMorse.forEach( (character, string) -> morseToChar.put(
      string, character ) );
  }

  public String encode( String string ) {
    StringJoiner result = new StringJoiner( " " );
    for ( int i = 0; i < string.length(); i++ ) {
      char c = string.charAt( i );
      if ( c == ' ' )
        result.add( "" );
      else {

        String maybeMorse = charToMorse.get( Character.toLowerCase( c ) );
        if ( maybeMorse != null )
          result.add( maybeMorse );
      }
    }
```

```java
    return result.toString();
  }

  public String decode( String string ) {
    StringBuilder result = new StringBuilder( string.length() / 4 );

    for ( String word : string.split( " {2}" ) ) {
      for ( Scanner scanner = new Scanner( word ); scanner.hasNext(); )
        Optional.of( scanner.next() ).map( morseToChar::get )
                .ifPresent( result::append );
      result.append( ' ' );
    }

    return result.toString();
  }
}
```
列表 3.28 com/tutego/exercise/util/MorseDemo.java

为了将字母映射到摩尔斯密码，我们使用 HashMap 类型的关联映射。不需要对键进行排序，因此不需要 TreeMap。代码量表明创建 Map 的工作不容易。

由于我们需要双向转换，所以使用两个映射。第一个 Map charToMorse 将字符与摩尔斯密码关联，第二个 Map morseToChar 将摩尔斯密码与字符关联。我们可以从第一个 Map 创建第二个。为此，运行第一个 Map，提取键和值并以相反的顺序放置在第二个 Map 中。

encode(String) 方法遍历字符串并将其转换为摩尔斯密码。结果是动态创建的，这实际上是 StringBuilder 的一个典型任务，但是这里使用了一个 StringJoiner。这个类很有用，因为它一方面是字符串的动态数据结构，另一方面会产生一个结果，其中每个子字符序列可以由用户定义的分隔符分隔，在我们的例子中是空格。

一个简单的 for 循环访问每个字符。如果字符是空格，那么我们将一个空字符串放入 StringJoiner，这会导致结果中有两个空格（为什么？因为 "+" "+"）。如果不考虑空格，那么我们将字符转换为小写，然后查询 charToMorse 关联映射。如果字符没有对应的摩尔斯密码，则 get(…) 方法将返回 null。在这种情况下，没有什么可做的。否则，我们将摩尔斯密码传递给 StringJoiner。

decode(String) 方法则相反。我们得到一长串摩尔斯密码序列，需要将其转换回原始文本。该方法的结果是使用 StringBuilder 动态创建的字符串。我们用一个容量初始化它并估计结果有多大；我们估计它将是输入字符串原始大小的四分之一。

摩尔斯单词被两个空格隔开，因此我们先询问单词。split("{2}") 给了我们所有单词；当然，带有两个空格的字符串也可以，但在代码中不会那么明显。

转换为摩尔斯密码的字母和数字用空格分隔。Scanner 帮助拆分，StringTokenizer 或 split(...) 也可以解决问题。使用提取的子字符串查询第二个数据结构 morseToChar。符号序列不存在，然后 Map morseToChar 上的 get(...) 将返回 null。使用 Optional 我们可以很好地表达这个级联：

1. 使用标记创建一个 Optional，这永远不会为 null。
2. 将标记映射到来自 morseToChar 的条目。如果映射没有关联的值，则 Optional 变为空。
3. 如果存在关联，则将关联的值附加到 StringBuilder。

当然，Optional 也可以与 encode(...) 一起使用，建议解决方案只使用了替代方案的一小部分。

最后，我们将 StringBuilder 转换为 String 并返回结果。

任务 3.4.3：用关联映射标记词频

```
public static final int LIMIT = 5;

public static List<String> importantGossip( String... words ) {

  Map<String, Integer> wordOccurrences = new HashMap<>( words.length );

  for ( String word : words )
    wordOccurrences.merge( word, 1, Integer::sum );

  Comparator<Map.Entry<String, Integer>> compareByWordOccurrence =
      Comparator.comparingInt( (ToIntFunction<Map.Entry<String, Integer>>)
        Map.Entry::getValue )
            .reversed()
            .thenComparing( Map.Entry::getKey );

  SortedSet<Map.Entry<String, Integer>> sortedSet =
    new TreeSet<>(compareByWordOccurrence );
  sortedSet.addAll( wordOccurrences.entrySet() );
```

```
      List<String> result = new ArrayList<>( LIMIT );
      for ( Map.Entry<String, Integer> element : sortedSet ) {
        result.add( element.getKey() );
        if ( result.size() >= LIMIT )
          break;
      }

      return result;
    }
```

列表 3.29 com/tutego/exercise/util/ImportantGossip.java

提出的解决方案分三步进行：第一步，计算文本中所有单词的频率；第二步，按频率排序；第三步，取出排序后的数据结构的前 5 个元素并准备返回。

为了记住频率，使用将字符串与整数（频率）关联的关联映射。通过循环，遍历所有单词并将它们放入 Map。可以使用 Java 8 中的一个快捷方法，即 merge(...) 方法：

```
default V merge(K key, V value, BiFunction<? super V, ? super V, ?
  extends V> remappingFunction)
```

我们可以使用它为新键设置初始值（此处为 1）或为现有键调用一个函数，将旧值连接到新值（在我们的例子中还是 1），然后将其写回。

循环运行后，每个单词及其频率都存储在 wordOccurrences 中。关联映射可以有两种不同的形式：作为 HashMap 或作为 TreeMap。无论我们使用什么，键均处于中心位置，相关的值处于次要位置。然而，在我们的例子中，关联值对于解决方案很重要。每个 Map 都提供了一个方法，我们可以使用它来一起读取键值对：entry Set( )。结果是一组 Map.Entry 对象。可以将这些 Map.Entry 对象放入一个新的数据结构，然后按照我们使用 getValue( ) 从 Map.Entry 获得的频率对它们进行排序。

建议解决方案使用填充了来自 wordOccurrences 的 Map.Entry 对象的 TreeSet。TreeSet 是一个有序集合，我们需要一个排序标准。Comparator compareByWordOccurrence 确定 Map.Entry 对象的顺序。这必须以 Map.Entry 对象的频率为第一标准，如果频率相同，则应添加单词的字典顺序作为比较。Comparator 使用静态和 default 方法很容易创建，它从 Comparator.comparingInt(...) 开始提取频率，并且由于需要高频词先出现，低频词后出现，所以我们必须颠倒默认顺序，reversed( ) 负责处理。如果一个频率出现两次，则使用 thenComparing(...) 添加第二个 Comparator，它使用单词的顺序作为第二个标准。

TreeSet 是使用 Comparator compareByWordOccurrence 创建的，所有 Map.Entry 对象都进入排序集合——TreeSet 将在添加时自动按频率对它们进行排序。现在我们只需要从排序集合中读出前 5 个元素。我们使用 Iterator 来做到这一点。这被我们用来开始遍历集合的扩展的 for 循环间接使用。每个元素都被放入我们的返回——ArrayList。当 ArrayList 为 5 个元素长时，我们中断循环并在最后返回 ArrayList。如果单词列表中的单词少于 5 个，则循环会预先中断，而不是通过条件判断。最后 ArrayList 中的元素少于 5 个。

任务 3.4.4：读取颜色并播放

```
public class ColorNames {
  public static class Color {
    private final String name;
    private final int rgb;

    private Color( String name, String rgb ) {
      this.name = Objects.requireNonNull( name );
      this.rgb = decodeHexRgb( rgb );
    }

    public static int decodeHexRgb( String hexRgb ) {
      if ( !hexRgb.startsWith( "#" ) )
        throw new IllegalArgumentException( "hex does not start with #" );
      if ( !(hexRgb.length() == 4 || hexRgb.length() == 7) )
        throw new IllegalArgumentException( hexRgb +
          " is not neither 4 (#RGB) nor 7 symbols (#RRGGBB) long" );

      if ( hexRgb.length() == 4 )
        hexRgb = "#" + hexRgb.charAt( 1 ) + hexRgb.charAt( 1 )
            + hexRgb.charAt( 2 ) + hexRgb.charAt( 2 )
            + hexRgb.charAt( 3 ) + hexRgb.charAt( 3 );
      return Integer.decode( hexRgb );
    }

    @Override public String toString() {
      return String.format( "'%s' is RGB #%06X", name, rgb );
    }
```

```java
    }

    private final HashMap<Integer, Color> colorMap = new HashMap<>();

    public ColorNames( String filename ) throws IOException {
      for ( String line : Files.readAllLines( Paths.get( filename ) ) ) {
        String[] tokens = line.split( "([\",]+" );
        Color color = new Color( tokens[ 1 ], tokens[ 2 ] );
        colorMap.put( color.rgb, color );
      }
    }

    public Optional<Color> decode( int rgb ) {
      return Optional.ofNullable( colorMap.get( rgb ) );
    }
  }
```

列表 3.30 com/tutego/exercise/util/ColorNames.java

在建议解决方案中，Color 类用作 ColorNames 中的嵌套静态类。这种关联是有意义的，因为 Color 对象仅在 ColorNames 的上下文中相关。Color 对象具有名称和 RGB 值所需的对象变量，并提供了一个获取名称和 RGB 值的构造函数。对象变量是私有的，构造函数也是如此——外部类仍然可以使用私有构造函数，这是 Java 可见性的一个特点。构造函数将 RGB 值作为字符串，这意味着它必须重新编码。Color 类为此声明了自己的方法。

int decodeHexRgb(String) 首先检查字符串的有效性。如果不以 # 开头，则不正确。即使字符串不是 4 或 7 个字符长，这也是一个错误，并且会出现 IllegalArgumentException。虽然文件中的 RGB 值是正确的，但是这个公共方法是对所有字符串都可用的，不正确的字符串应该注意。如果字符串只包含 4 个符号，那么就是简写，红、绿、蓝的值加倍。Integer.decode(...) 将相应的 RGB 值作为整数返回。

ColorNames 也有一个构造函数，它接收文件名。Files.readAllLines(...) 读取文件的所有行，扩展的 for 循环逐行遍历它们。使用 split(...) 方法将每行标记为字符串，我们访问数组中的第二个和第三个元素——即索引 1 和 2——并用它们填充构造函数。我们将生成的 Color 对象放入 Map，将整数形式的 RGB 值作为与其关联的 Color 对象的键。

剩下的方法 decode(String) 询问 Map，如果 RGB 值和颜色之间没有关联，则可能返回 null。由于我们要避免返回 null，所以 Optional.ofNullable(...) 帮助未知的 RGB 值变成 Optional.empty( )。

任务 3.4.5：读取名称，管理长度

```java
SortedMap<Integer, List<String>> namesByLength = new TreeMap<>();

InputStream resource = FamilyNamesByLength.class.getResourceAsStream(
    "family-names.txt" );
try ( Scanner scanner = new Scanner( resource ) ) {
  while ( scanner.hasNextLine() ) {
    String name = scanner.nextLine();
    namesByLength.computeIfAbsent( name.length(), __ -> new ArrayList<>() )
                 .add( name );
  }
}

namesByLength.forEach( (len, names) -> System.out.println( len + " " +
  names ) );

for ( int len; (len = new Scanner( System.in ).nextInt()) > 0; ) {
  int finalLen = len;
  Optional.ofNullable( namesByLength.get( len ) )
          .ifPresentOrElse(
             System.out::println,
             () -> System.out.printf(
                "No words of length%d%n", finalLen ) );
}
```

列表 3.31 com/tutego/exercise/util/FamilyNamesByLength.java

SortedMap<Integer, List<String>> 类型表示整数与列表的关联。在字符串可以被包含在列表中之前，必须创建列表。程序必须执行以下操作：

1. 它必须检查是否已经有一个长度列表，如果有，则可以追加字符串。
2. 如果长度在 Map 中还没有条目，则必须将一个列表与该长度关联，并且该长度的第一个单词必须被包含在该列表中。

在代码中，这可以表示如下：

```
if ( ! namesByLength.containsKey( line.length() ) )
  namesByLength.put( line.length(), new ArrayList<>() );

namesByLength.get( line.length() ).add( line );
```

它测试 containsKey(...) 中是否已经存在该字符串长度的列表；如果不存在，则为该字符串长度创建一个新的 ArrayList。在下一个语句中添加 put(...) 始终会成功，因为保证了一个长度的列表。

从 Java 8 开始，我们不再需要手动编写它，而是可以使用默认的 computeIfAbsent(...) 方法。描述有点复杂，从源码看方法更容易理解：

```
default V computeIfAbsent( K key,
    Function<? super K, ? extends V> mappingFunction ) {
  Objects.requireNonNull( mappingFunction );
  V v;
  if ( (v = get( key )) == null ) {
    V newValue;
    if ( (newValue = mappingFunction.apply( key )) != null ) {
      put( key, newValue );
      return newValue;
    }
  }

  return v;
}
```

列表 3.32 Aus der OpenJDK-Implementierung von java.util.Map

首先调用 get(...) 方法，有两个出口：

1. 如果值不为 null，则直接返回。在我们的例子中，这意味着已经建立了一个字符串长度列表。
2. get(...) 在没有关联的情况下返回 null（我们忽略了键可能与 null 关联）。如果没有关联值，则使用键调用传递的函数。该函数返回一个值，并且该值存储在键下，至少当它不为 null 时。在我们的例子中，该函数创建了一个新的 ArrayList，并且键的名称是不必要的。因此，程序用两个下划线隐藏了这个标识符。给读者的问题：与 __-> new ArrayList<>( ) 相比，有什么理由反对构造函数引用 ArrayList::new？它也具有相同的功能。

computeIfAbsent(...) 在任何情况下都返回最后的列表，我们可以级联 add(...)。

在程序逐行读取文件并填充数据结构后，Map 类的 forEach(BiConsumer) 方法将遍历所有条目。我们的 BiConsumer 获取键和值并在屏幕上输出键值对。

程序的最后一部分在一个循环中，只有当输入为 0 或负数时才退出。程序要求输入一个整数作为词长，然后 get(...) 请求列表，但如果没有要输入的列表，则结果可能为 null。Optional.ofNullable(...) 很好地将可能的 null 包装在 Optional 中，因为如果没有与键关联的值，则 Optional.isEmpty( ) 为真。ifPresentOrElse(...) 就像条件判断：如果 Optional 包含值，则我们的 Consumer 输出姓名列表；如果没有关联值，则屏幕上会显示一条消息，指出不存在所需长度的单词。赋值技巧 int finalLen = len 是必要的，因为 Lambda 表达式只能访问最终变量，但计数器 len 会发生变化。将其复制到中间变量可以解决该问题。

任务 3.4.6：找到缺失的字符

该解决方案由两部分组成：首先建立一个对我们来说是最优的数据结构，然后查询该数据结构。

```
List<String> words = Arrays.asList( "haus", "maus", "elefant",
"klein", "groß" );

Map<String, List<String>> map = new HashMap<>();

// Initialize the map
for ( String word : words ) {
  for ( int index = 0; index < word.length(); index++ )
    map.computeIfAbsent( index + "-" + word.charAt( index ),
                        __ -> new ArrayList<>() ).add( word );
}
```

列表 3.33 com/tutego/exercise/util/FindMissingLetters.java

该任务可以通过多种数据结构和方法来完成。

此处选择的版本执行以下操作：创建了一个 Map<String, List<String>>，其中字母的位置与单词列表关联。我们以 haus（房子）和 maus（老鼠）为例。这两个词在索引 1 处都有一个 a。索引和字母之间的这种联系是关联映射的关键。

程序应将输入 "haus" "maus" "elefant" "klein" "groß" 转换为以下 Map：

{3-ß=[groß], 1-a=[haus, maus], 4-a=[elefant], 0-e=[elefant], 0-g=[groß], 2-e=[elefant, klein], 0-h=[haus], 3-f=[elefant], 0-k=[klein], 3-i=[klein], 1-l=[elefant, klein], 0-m=[maus], 2-o=[groß], 4-n=[klein], 1-r=[groß], 5-n=[elefant], 3-s=[haus, maus], 2-u=[haus, maus], 6-t=[elefant]}

为了创建 Map，外循环遍历所有单词，然后内循环遍历所选单词的每个字母。索引中的对、减号和位置进入关联映射。由于关联映射不直接将这个键与单词联系起来，而是将字符串放在一个列表中，所以每次将第一个单词放入列表时都必须重新创建该列表。computeIfAbsent(K key,Function mappingFunction) 非常适合此类任务；如果键下没有关联列表，则调用映射函数并返回一个新列表，然后与该键关联。computeIfAbsent(...) 方法的结果是关联值，也就是我们可以在其中放置单词的列表。

第二部分可以查询数据结构，解决未知字符的问题。

```java
// Query the data structure
Consumer<String> letterFinder = word -> {
  Set<String> matches = null;
  for ( int index = 0; index < word.length(); index++ ) {
    // Skip unknown chars
    if ( word.charAt( index ) == '_' )
      continue;
    List<String> wordCandidates = map.get(
      index + "-" + word.charAt( index ) );
    // Exit loop if no known entry
    if ( wordCandidates == null ) {
      // Remove possible previous matches for correct console output
      matches = null;
      break;
    }
    // Build a copy and remove words that don't match with the length
    wordCandidates = new ArrayList<>( wordCandidates );
    wordCandidates.removeIf( s -> s.length() != word.length() );
    // Join matches from all known letters
```

```
    if ( matches == null )
      matches = new HashSet<>( wordCandidates );
    else
      matches.retainAll( wordCandidates );
  }
  System.out.println( word + " -> " +
    (matches == null || matches.isEmpty() ? "No results" : matches) );
};

List<String> missingLettersWords =
  Arrays.asList( "ha__", "el__a_t", "x", "hi__", "___s", "___s___" );
missingLettersWords.forEach( letterFinder );
```

列表 3.34 com/tutego/exercise/util/FindMissingLetters.java

Consumer 是一种子程序，它获取一个带下划线的单词，然后查询数据结构 Map 并将匹配的单词输出到屏幕上。匹配的词被收集在集合 matches 中。该算法的流程如下：从每个字符中找出在同一地方有相同字符的所有可能的词，把它们放在集合中，随后与添加在后面的字符形成交集。

假如一开始有很多第一位是 h 的候选词，但加上第二位限定为 a 的话，候选词数量会减少。

在程序中可设置一个大主循环，从头到尾运行源单词，并请求位置 0、位置 1 等的所有字符。未知字符对识别没有帮助（尽管我们可以接受所有单词），因此程序返回循环。如果字符不是下划线，则我们用索引、减号和字符生成一个键。我们使用该键查询 Map，最好的情况是得到一个候选列表。但是，可能没有该键的列表，即源词中有一个字符，但不匹配。没有解决方案，集合 matches 变为 null，我们可以退出循环。

如果我们找到了与字符匹配的单词列表，则程序首先创建列表的副本，然后删除所有与源单词长度不同的单词。现在必须把先前的单词和 index 中的新单词进行交集，因为结果必须出现在两个集合中。如果之前没有找到任何单词，则 matches 为 null，我们使用当前候选词创建一个新集合。如果已经有结果并且 matches 不为 null，则 retainAll(...) 从 matches 中删除所有未出现在 wordCandidates 中的单词。

循环结束时，我们运行了所有已知字符，不断减小集合 matches 的大小。控制台输出向我们显示该集合的内容，如果该集合为空，则显示一条信息。

任务 3.4.7：计算寻找三头猴的路径数

用于缓存的数据结构 WeakHashMap 使用了所谓的弱引用。Java 中基本上有四种类型的引用：

- 强引用是我们作为 Java 程序员每天处理的常见引用。垃圾收集器永远不会解析这些引用。
- WeakReference（弱引用）是在垃圾收集器运行时解析的引用。确切的时间取决于 JVM 的实现。
- 使用 SoftReference（软引用）时，JVM 会尝试让对象保持活动状态，直到出现 OutOfMemoryError。
- 对于 PhantomReference（虚引用），我们只知道垃圾收集器已经移除了对象。

WeakHashMap 在内部使用 WeakReference，这样我们不必直接处理这些引用或清理垃圾收集器。它是一个可用的数据结构，只要有足够的空闲内存就会保留引用，然后在垃圾收集器需要腾出空间时释放对象。对于我们的计算来说，这是一个很好的选择，因为 WeakHashMap 会在计算阶乘时被填满，但这些阶乘值不需要长时间保存在内存中，而是随后会被清除。

WeakHashMap 是一个特殊的 Map，这意味着我们可以使用所有已知的方法（见图 3.3）。

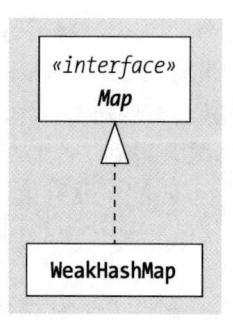

图 3.3 实现映射的 WeakHashMap 的 UML 图示

```
private static final Map<BigInteger, BigInteger> factorialCache =
  new WeakHashMap<>();

private static BigInteger factorial( BigInteger n ) {
  BigInteger maybeCachedValue = factorialCache.get( n );
```

```
    if ( maybeCachedValue != null )
      return maybeCachedValue;

    // n < 2 ? 1 : n * factorial( n - 1 )
    BigInteger result = isLessThan( n, TWO )
      ? ONE
      : n.multiply( factorial( n.subtract( ONE ) ) );

    factorialCache.put( n, result );
    return result;
  }

  private static boolean isLessThan( BigInteger a, BigInteger b ) {
    return a.compareTo( b ) < 0;
  }

  public static BigInteger catalan( BigInteger n ) {
    // (2n)! / (n+1)!n!
    BigInteger numerator   = factorial( TWO.multiply( n ) );
    BigInteger denominator = factorial( n.add( ONE ) ).multiply(
      factorial( n ) );
    return numerator.divide( denominator );
  }
```

列表 3.35 com/tutego/exercise/util/CachedCatalan.java

factorial(...) 是需要访问缓存的方法。因此，我们首先为内部缓存创建一个对象变量 factorialCache，它可以保存 BigInteger。关联是从 n 到 n! 的映射。两种类型都是 BigInteger。

使用缓存包括两个主要步骤：

1. 请求值时，首先询问缓存是否包含该值。如果它在缓存中，那么我们将很快完成。
2. 如果该值不在缓存中，则计算并放入缓存。这需要一段时间。

factorial(...) 方法就是这么实现的。查询缓存，可能有结果，也可能没有结果。我们已经知道 Map 的 get(...) 方法的响应行为，如果没有关联值，则该方法返回

null。指示器说明我们是否在缓存中有值。如果结果不是 null，那么我们将缓存中的 BigInteger 作为计算的阶乘，并可以直接返回它。但是，如果 get(...) 返回 null，那么我们必须计算阶乘，然后将结果放入 factorialCache 并返回。BigInteger 中没有专门的方法来比较一个 BigInteger 比另一个 BigInteger 大还是小，因为 BigInteger 具有自然顺序，因此实现了 Comparable。但是，compareTo(...) 的可读性不是最佳的，因此引入了一个单独的方法 isLessThan(...)。BigInteger 常量 ONE 和 TWO 已被静态导入，这稍微缩短了代码。

另一种实现可以使用 computeIfAbsent(...)，但这里提供的代码很容易理解。

catalan(...) 方法执行记录的计算并调用 factorial(...) 三次。许多条目已经在缓存中，因此 factorial(...) 可以提供来自缓存的许多响应。

任务 3.4.8：在排序的关联映射中管理节日

```
NavigableMap<LocalDate, String> dates = new TreeMap<>();
    dates.put( LocalDate.of( 2020, Month.JANUARY, 1 ), "Neujahr" );
    dates.put( LocalDate.of( 2020, Month.APRIL, 10 ), "Karfreitag" );
    dates.put( LocalDate.of( 2020, Month.MAY, 1 ), "Tag der Arbeit" );
    dates.put( LocalDate.of( 2020, Month.JUNE, 1 ), "Pfingstmontag" );
    dates.put( LocalDate.of( 2020, Month.OCTOBER, 3 ), "Tag der Deutschen Einheit" );
    dates.put( LocalDate.of( 2020, Month.NOVEMBER, 1 ), "Allerheiligen" );
    dates.put( LocalDate.of( 2020, Month.JULY, 1 ), "Pfingstmontag" );
    dates.put( LocalDate.of( 2020, Month.DECEMBER, 25 ), "1. Weihnachtsfeiertag" );
    dates.put( LocalDate.of( 2020, Month.DECEMBER, 26 ), "2. Weihnachtsfeiertag" );
    dates.put( LocalDate.of( 2021, Month.JANUARY, 1 ), "Neujahr" );
    dates.put( LocalDate.of( 2021, Month.APRIL, 20 ), "Karfreitag" );

    System.out.println( dates.firstEntry() );
    System.out.println( dates.lastEntry() );

LocalDate festiveSeasonStart = LocalDate.of( 2020, Month.DECEMBER, 23 );
LocalDate festiveSeasonEnd   = LocalDate.of( 2021, Month.JANUARY, 6 );

    System.out.println( dates.higherEntry( festiveSeasonEnd ) );
```

```
SortedMap<LocalDate, String> festiveSeason =
    dates.subMap( festiveSeasonStart, true, festiveSeasonEnd, true );

System.out.printf( "%d Feiertage:%n", festiveSeason.size() );
festiveSeason.forEach(
  ( date, name ) -> System.out.printf( "%s am%s%n", name, date ) );

festiveSeason.clear();

System.out.println( dates );
```

列表 3.36 com/tutego/exercise/util/Holidays.java

（注：Neujahr——新年、Karfreitag——耶稣受难日、Tag der Arbeit——劳动节、Pfingstmontag——圣灵降临节后的星期一、Tag der deutschen Einheit——德国国庆节、Allerheiligen——万圣节、1. Weihnachtsfeiertag——圣诞节第一天、2. Weihnachtsfeiertag——圣诞节第二天。）

建议解决方案使用所描述的方法 firstEntry( )、lastEntry( )、higherEntry(K) 和 subMap(K, boolean, K, boolean)。

NavigableMap 的不同方法返回所谓的视图（英语: views）。视图不是数据的副本，但对该视图的操作总是会进入底层数据结构。我们可以在视图上使用 clear( ) 方法来实现部分区域的删除。视图可以节省内存并且性能良好，但可能导致错误，因为我们可能在认为自己在处理副本时不小心修改了底层数据结构。当底层数据结构是不可变的，即根本无法更改，但元素被插入视图，它可能导致问题。

测试 3.4.9: HashMap 中的值
程序输出如下：

```
{java.awt.Point[x=2,y=1]=java.awt.Point[x=1,y=2]}
null
java.awt.Point[x=1,y=2]
```

插入的值对象必须是不可变的，否则动态计算的哈希码会发生变化，无法再找到对象。

任务 3.4.10: 确定共同点：派对场所布置和带来的礼物
```
private static void printMultipleGifts( List<Set<String>> families ) {
```

```java
class Bag extends HashMap<String, Integer> {
  void add( String key ) { merge( key, 1, Integer::sum ); }
  // int getCount( String key ) { return getOrDefault( key, 0 ); }
}

Bag giftsToCounter = new Bag();

for ( Set<String> gifts : families )
  for ( String gift : gifts )
    giftsToCounter.add( gift );

System.out.println( giftsToCounter );

giftsToCounter.forEach( (gift, counter) -> {
  if ( counter > 1 )
    System.out.println( gift );
} );
}
```

列表 3.37 com/tutego/exercise/util/CommonGifts.java

我们将相应礼物的计数和发放交给 printMultipleGifts(...) 方法。它得到一份集合列表，每份都是一个家庭带来的礼物。计算相同的事物是一项常见任务，但无法使用 Java 中的内置数据结构来实现。因此，我们创建了一个扩展 HashMap 的小型本地类 Bag。这个 Bag 类将一个字符串（即礼物）与频率的整数关联。我们给这个小类一个方法，这样就可以从外部增加一个键的值；在内部，关联值增加 1。这里我们使用 Map 提供的 merge(...) 方法。如果没有值与键关联，则设置 1，如果至少有一个值，则将 1 添加到旧值并更新数据记录。

我们的方法创建了一个 Bag 实例，遍历所有带来礼物的家庭集合，然后遍历所有礼物本身，并为每个礼物调用 add(...) 方法。

如果我们想知道哪个礼物出现了不止一次，则可以使用 Map 的 forEach(...) 方法遍历所有键值对并询问计数器的计数值是否大于 1，如果计数器的计数值大于 1，则输出礼物。

任务 3.5.1：开发便捷的属性装饰器

```java
public class PropertiesConfiguration {
  private final Properties properties;
```

```java
  public PropertiesConfiguration( Properties properties ) {
    this.properties = properties;
  }

  public Optional<String> getString( String key ) {
    return Optional.ofNullable( properties.getProperty( key ) );
  }

  public Optional<Boolean> getBoolean( String key ) {
    try { return getString( key ).map( Boolean::valueOf ); }
    catch ( Exception e ) { return Optional.empty(); }
  }

  public Optional<BigInteger> getBigInteger( String key ) {
    try { return getString( key ).map( BigInteger::new ); }
    catch ( Exception e ) { return Optional.empty(); }
  }

  public OptionalDouble getDouble( String key ) {
    try { return OptionalDouble.of( Double.parseDouble(
      properties.getProperty( key ) ) ); }
    catch ( Exception e ) { return OptionalDouble.empty(); }
}

  public OptionalLong getLong( String key ) {
    try { return OptionalLong.of( Long.parseLong(
      properties.getProperty( key ) ) ); }
    catch ( Exception e ) { return OptionalLong.empty(); }
  }

  public List<String> getList( String key ) {
    List<String> result = getString( key )
      .map( s -> s.split( "\\s*(?<!\\\\),\\s*" ) )
      .map( Arrays::asList )
      .orElse( Collections.emptyList() );
    result.replaceAll( string -> string.replace( "\\,", "," ) );
```

```
      return result;
    }

    public void putBinary( String key, byte[] bytes ) {
      String base64 = Base64.getEncoder().encodeToString( bytes );
      properties.setProperty( key, base64 );
    }

    public Optional<byte[]> getBinary( String key ) {
      return getString( key ).map(
        base64 -> Base64.getDecoder().decode( base64 ) );
    }
  }
```

列表 3.38 com/tutego/exercise/util/PropertiesConfiguration.java

PropertiesConfiguration 类有一个接受和存储实际 Properties 的构造函数。我们的方法比 Properties 提供的方法更灵活，但我们的方法在内部基于 Properties 类的方法。这意味着我们提供的所有方法都以某种方式使用了 Properties 类的方法。重点是 getProperty(String) 方法，我们也使用 setProperty(...) 来设置属性。

- Optional<String> getString(String)。属性可能存在，也可能不存在。如果我们使用简单方法 getProperty(String) 找到一个不存在的属性，则 getProperty(...) 将返回 null。虽然有一个 getProperty(String key, String defaultValue) 方法，但它不会阻止返回 null 或要求调用者进行 null 处理。使用 Optional 数据类型，Java 提供了一种很好的方式来表示所请求的值不存在。此外，Optional 为我们提供了一个很好的机会来制定替代方案，如使用 Optional 方法 orElse(T other) 或 orElseGet(Supplier<? extends T> other)，甚至 orElseThrow(Supplier<? extends X> exceptionSupplier)。

- Optional<Boolean> getBoolean(String) 和 Optional<BigInteger> getBigInteger(String)。实现清楚地表明 Optional 数据类型有一个 map(Function) 方法，我们可以使用该方法将字符串转换为数据类型 Boolean 和 BigInteger 并打包在 Optional 中。但是，由于基础 Properties 对象中的格式可能是错误的，所以在转换过程中可能发生异常。我们拦截异常，在这种情况下，我们的方法返回 Optional.empty( )。

- OptionalDouble getDouble(String), OptionalLong getLong(String key)。对于基本数据类型 double 和 long，有特殊的数据类型 Optional，即 OptionalDouble

和 OptionalLong。其逻辑类似。
- List<String> getList(String)。如果我们返回一个列表，则使用我们自己的 getString(...)，因为它提供了一个 Optional，我们可以直接在此设置一个 map(...)，将用逗号分隔的各条目转换为一个字符串数组。map(...) 的下一次调用将数组转换为列表。如果源 getString(...) 返回一个 Optional.empty()，则 orElse(...) 将返回一个空列表。我们约定，逗号必须用 "\" 转义，然后不被视为分隔符。其他字符串仍然包含 "\"，而不是 ","。我们使用 List 方法 replaceAll(...) 解决这个问题，该方法对每个元素执行替换功能。
- void putBinary(String, byte[ ]), Optional<byte[ ]> getBinary(String)。前面的方法只是方便阅读的方法。使用 putBinary(...)，我们有一个 byte 数组 Base64 进行编码并将其放入 Properties 对象的方法。Java 为 Base64 转换提供了 Base64 类。getEncoder() 和 getDecoder() 方法返回用于将 byte 数组转换为字符串并将字符串转换为 byte 数组的对象。我们的 getBinary(...) 方法将字符串转换为数组。同样，我们自己的 getString(...) 方法返回一个 Optional，map(...) 方法将 String 转换为 byte[]。

任务 3.6.1：编辑 RPN（逆波兰表示法）计算器

```
String input = "160 50 30 + /";
Queue<Integer> stack = Collections.asLifoQueue( new ArrayDeque<>() );

Pattern operatorPattern = Pattern.compile( "[+*/-]" );
Pattern numericPattern = Pattern.compile( "\\d+" );
for ( String token : input.split( "\\s+" ) ) {
  Matcher operatorMatcher = operatorPattern.matcher( token );
  Matcher numericMatcher = numericPattern.matcher( token );
  if ( numericMatcher.matches() )
    stack.add( Integer.parseInt( token ) );
  else if ( operatorMatcher.matches() ) {
    int operand2 = stack.remove();
    int operand1 = stack.remove();
    switch ( token ) {
      case "+": stack.add( operand1 + operand2 ); break;
      case "-": stack.add( operand1 - operand2 ); break;
      case "*": stack.add( operand1 * operand2 ); break;
      case "/": stack.add( operand1 / operand2 ); break;
    }
```

```
    }
    else
      System.out.println( "Unknown type!" );
}
System.out.printf( "Result:%d", stack.remove() );
```
列表 3.39 com/tutego/exercise/util/UPN.java

RPN 计算器分两步工作：

1. 将字符串拆分为标记；
2. 处理标记（这里我们区分标记是数字还是符号）。

  实际的算法可以以不同的方式实现。一种方法是首先将所有数字收集在一个堆栈上，并将所有运算符收集在第二个堆栈上。最后，使用值的堆栈一起处理带有运算符的堆栈。我们使用一种不同的方法：只有一个堆栈。
  Java 提供了 java.util.Stack 数据结构，但该数据结构是 Java 1.0 已弃用的类型之一，不应再使用。因此，我们使用 LIFO 队列，add(...) 在后面添加一些东西（后入），remove(...) 从队列中删除一些东西（先出）。
  如果标记是整数的字符串表示，那么我们将字符串转换为整数并将其放入堆栈。我们还使用正则表达式来检查标记是否为二元运算符。在对正则表达式 [+*/–] 中的运算符进行排序时，注意符号之间不要出现"–"，否则该减号表示 a-z 之类的范围规范，这当然是我们不想要的。
  如果标记是运算符，则需要从堆栈中取两个操作数。重要的是要注意先从堆栈中取出第二个操作数，然后取出第一个操作数。加法和乘法是可交换的，但减法和除法不交换。switch 语句帮助我们对标记执行正确的操作并将结果放回堆栈。
  由于二元运算总是由三个符号组成，所以在解析后只剩下一个数字。一旦我们处理了输入中的所有标记并且数字和运算符的数量已经平衡，最后就只有一个数字，即结果。读者应考虑到不正确的输入值可能导致出现什么错误。

任务 3.7.1：查找重复条目并解决动物混乱

```java
private static String sameSymbols( String string1, String string2 ) {

  BitSet bits = new BitSet( 1024 );

  string1.codePoints().forEach( character -> bits.set( character ) );
```

```java
    int capacity = (int) ((long) string1.length() + string2.length()) / 2;
    StringBuilder result = new StringBuilder( capacity );
    string2.codePoints().forEach( character -> {
      if ( bits.get( character ) )
        result.appendCodePoint( character );
    } );

    return result.toString();
}
```

列表 3.40 com/tutego/exercise/util/AnimalMissing.java

这种任务完成方法如下：从第一个字符串中取出第一个字符，再看这个字符是否出现在第二个字符串中。如果出现，则标记这个字符，然后从第一个字符串中取出下一个字符并再次测试该字符是否出现在第二个字符串中。然而，问题在于运行时间将是二次幂的，因为我们必须一遍又一遍地从头到尾遍历第二个字符串。如果想节省时间，则必须用不同的方式完成任务。

解决方案如下：记住第一个字符串中出现的每个字符。我们可以使用一个集合，即 Set<Character>，但这里有一个替代解决方案。BitSet 是一种特殊的关联映射，它将整数与 boolean 关联。整数不应太大，否则 BitSet 将需要更多内存，但由于 Unicode 字符的大小有限制，所以内存需求保持可控。我们可以很容易地计算出来。Unicode 12 大约有 140 000 个字符，这就是我们需要的位数。140 000 位 / 8 = 17 500 字节，即 17 KB，相当于一张小图片的大小。

语句 string1.codePoints( ).forEach(...) 提取第一个字符串的每个字符。每个字符由一个代码点表示。代码点变成一个索引，此时我们在 BitSet 中设置一个位来标记它。codePoints( ) 返回一个 IntStream，这是一种特殊的数据类型，将在下一章更深入地探讨。

我们对第二个字符串进行类似的处理。这里也得到一个字符的 IntStream，但我们没有设置位，而是询问是否在位置 character 的 BitSet 中设置了位。如果是这样，那么相同的字符出现在第一个和第二个字符串中。我们想记住 StringBuilder 中的字符。StringBuilder 的大小被估计为 string1 和 string2 两个长度的算术平均值，我们必须在 long 类型的值范围中进行加法运算，以免在两个 int 值相加时出现溢出的风险。

遍历之后，我们将 StringBuilder 转换为 String 并返回它。运行时间是线性的，取决于两个字符串的大小，因为我们只需要运行第一个字符串一次，然后运行第二个字符串一次。

任务 3.8.1：装船

建议解决方案由三部分组成：Loader 和 Unloader 的声明以及线程的启动。

数据交换通过 java.util.concurrent.BlockingQueue 实现 ArrayBlockingQueue。我们的应用程序需要封锁，因此使用 put(...) 和 take()。

```java
class Loader implements Runnable {
  private final BlockingQueue<String> ramp;

  Loader( BlockingQueue<String> ramp ) {
    this.ramp = ramp;
  }

  @Override
  public void run() {
    while ( ! Thread.currentThread().isInterrupted() ) {
      try {
        String[] products = { "Rum", "wine", "salami", "beer", "cheese",
           "comics" };
        String product =
          products[ ThreadLocalRandom.current().nextInt( products.length ) ]
          + ":" + UUID.randomUUID().toString();
        ramp.put( product );
        System.out.printf( "Product with ID%s placed on the ramp%n",
product );
        TimeUnit.MILLISECONDS.sleep( ThreadLocalRandom.current().nextInt(
          1000, 2000 ) );
      }
      catch ( InterruptedException e ) { Thread.currentThread().
interrupt(); }
    }
  }
}
```

列表 3.41 com/tutego/exercise/thread/LoadingShips.java

Loader 是 Runnable，并且具有所需的构造函数，它采用 BlockingQueue，即 Loader 和 Unloader 稍后将用于交换数据的数据结构。run() 方法包含一个循环，只要外部发送中断，该循环就会终止。在我们的例子中虽然没有，但其实这是最佳实践。

在循环体中，随机选择一个产品，并生成一个带有随机标识符的产品名称。将该产品放置在坡道上。put(...) 方法会阻塞，因为坡道可能已经满了。如果线程从 put(...) 返回，则产品已被成功放置在坡道上。最后，线程将延迟执行几毫秒并继续循环。

```
class Unloader implements Runnable {
  private final BlockingQueue<String> ramp;

  Unloader( BlockingQueue<String> ramp ) {
    this.ramp = ramp;
  }

  @Override
  public void run() {
    while ( ! Thread.currentThread().isInterrupted() ) {
      try {
        String product = ramp.take();
        System.out.printf( "Product with ID%s taken off the ramp%n", product );
        TimeUnit.MILLISECONDS.sleep( ThreadLocalRandom.current().nextInt(
          1000, 2000 ) );
      }
      catch ( InterruptedException e ) { Thread.currentThread().interrupt(); }
    }
  }
}
```

列表 3.42 com/tutego/exercise/thread/LoadingShips.java

Unloader 类的创建类似。唯一的区别在于 while 循环的主体需要将产品从坡道中取出。我们使用 take() 方法。由于坡道上没有产品，所以 take() 方法可能阻塞。如果 take() 方法返回，则可以从坡道中取出一个产品并进行屏幕输出，同时程序运行会稍微延迟。

第 3 章 数据结构和算法 | 145

```java
class Unloader implements Runnable {
  private final BlockingQueue<String> ramp;

  Unloader( BlockingQueue<String> ramp ) {
    this.ramp = ramp;
  }

int RAMP_CAPACITY = 10;
BlockingQueue<String> ramp = new ArrayBlockingQueue<>( RAMP_CAPACITY );
ExecutorService executors = Executors.newCachedThreadPool();

for ( int i = 0;i< 5; i++ ) executors.execute( new Unloader( ramp ) );
for ( int i = 0;i< 10; i++ ) executors.execute( new Loader( ramp ) );
```
列表 3.43 com/tutego/exercise/thread/LoadingShips.java

最后一部分创建 ArrayBlockingQueue，传递 Loader 和 Unloader 这两个类，然后启动线程。

任务 3.8.2：优先编辑重要消息

```java
class Message {

  public final String message;
  public final long timestamp;

  Message( String message ) {
    this.message = Objects.requireNonNull( message );
    this.timestamp = System.nanoTime();
  }

  @Override public boolean equals( Object other ) {
    if ( other == null || getClass() != other.getClass() ) return false;
    return message.equals( ((Message) other).message ) &&
        ((Message) other).timestamp == timestamp;
  }

  @Override public int hashCode() {
```

```
      return Objects.hash( message, timestamp );
    }

    @Override public String toString() {
      return "'" + message + "\" + ", " + timestamp% 100_000;
    }
  }
```

列表 3.44 com/tutego/exercise/util/UrgentMessagesFirst.java

Message 类有一个构造函数，它接收并存储实际的消息。时间戳来自 System.nanoTime( )。equals(Object) 方法比较消息字符串和时间戳。hashCode( ) 是为了完整性而实现的，在我们的程序中没有调用它。toString( ) 将消息放在单引号中并使用时间戳的最后 5 位数字，因为这足以快速直观地比较哪个消息更新或更旧。

```
String KEYWORD = "Kanönchen";

Comparator<Message> keywordComparator = ( msg1, msg2 ) -> {
  boolean msg1HasKeyword = msg1.message.contains( KEYWORD );
  boolean msg2HasKeyword = msg2.message.contains( KEYWORD );
  boolean bothMessagesHaveKeywordOrNot = msg1HasKeyword == msg2HasKeyword;
  return bothMessagesHaveKeywordOrNot ? 0 : msg1HasKeyword ? -1 : +1;
};

Comparator<Message> messageComparator =
    keywordComparator.thenComparingLong( message -> message.timestamp );
PriorityQueue<Message> tasks = new PriorityQueue<>( messageComparator );
```

列表 3.45 com/tutego/exercise/util/UrgentMessagesFirst.java

keywordComparator 是解决方案的核心。它从 Message 中获取两条消息，并检查它们是否包含昵称。为了使代码更清晰，将 contains(...) 的结果缓存在两个变量中。现在有四种情况：两条消息都包含昵称，两条消息都不包含昵称，或者两条消息之一包含搜索的单词。如果两者都包含或不包含昵称，则 Comparator 返回 0，否则只有两个消息中的一条包含该字符串。如果昵称在第一条消息中，则根据 Comparator，该消息较小，并且 Comparator 以负值响应。否则，第二条消息中必须出现昵称，并且 Comparator 以正值响应。

如果 keywordComparator 返回 0，则 keywordComparator 后面的 Comparator 是必

要的。在这种情况下，时间应该作为第二个标准来考虑。设置的 thenComparing Long(...) 附加一个 Comparator，将时间戳考虑在内。该顺序是正确的，因为较旧的消息带有较小的时间戳。

将生成的 messageComparator 传递给 PriorityQueue 的构造函数，并初始化数据结构。

任务 3.8.3：用完就换新的

```java
class SecureRandomBigIntegerIterator implements Iterator<BigInteger> {

  private final SynchronousQueue<BigInteger> channel =
    new SynchronousQueue<>();

  public SecureRandomBigIntegerIterator() {
    Runnable bigIntegerPutter = () -> {
      try {
        while ( true ) {
          BigInteger bigInteger = internalNext();
          System.out.printf( "> About to put number%s... into the queue%n",
            bigInteger.toString().subSequence( 0, 20 ) );
          channel.put( bigInteger );
          System.out.println( "> Number was taken" );
        }
      }
       catch ( InterruptedException e ) { throw new IllegalStateException(e); }
    };
    ForkJoinPool.commonPool().submit( bigIntegerPutter );
  }

  private BigInteger internalNext() {
    return new BigInteger( 1024, new SecureRandom() );
  }

  @Override public boolean hasNext() {
    return true;
  }
```

```java
@Override public BigInteger next() {
  try {
    System.out.println( "< About to take a number" );
    BigInteger bigInteger = channel.take();
    System.out.println( "< Took a number out" );
    return bigInteger;
  }
  catch ( InterruptedException e ) { throw new IllegalStateException(e); }
}
```

列表 3.46 com/tutego/exercise/thread/SecureRandomBigIntegerIteratorDemo.java

解决方案的核心是 SynchronousQueue 类。它与普通 Queue 不同，本质上只用于将元素从一个线程传输到另一个线程。正是在这种情况下，我们使用了 SynchronousQueue。

我们的 SecureRandomBigIntegerIterator 类实现了 Iterator 接口并且可以无线返回许多数量的大整数。我们的类的构造函数创建一个 Runnable 并将其传递给预先配置的线程池，即 ForkJoinPool.commonPool( )。与自己的线程池相比，Fork-Join 池会在应用程序终止时自动关闭。这个特殊的池对于解决方案并不是真正必要的。

Runnable 包含一个无限循环，该循环立即开始请求整数。随机 BigInteger 来自内部方法。也就是说，即使在 Iterator 上根本没有调用 next( ) 方法，我们的实现也会继续并用 put(...) 方法填充 SynchronousQueue。在 put(...) 中，线程一直停留，直到对方从 SynchronousQueue 中取出元素，然后线程在这里继续无限循环并确定下一个数字。

Iterator 实例的 next(...) 方法从 SynchronousQueue 中获取数字，我们为这种情况创建了这个构造。使用 take( ) 方法获取元素，另外这会导致创建一个新的 BigInteger 并将其放入队列。如果在调用 next(...) 方法之间总是间隔一些时间，则后台的线程将已经计算出一个新的随机数，这意味着 next( ) 方法可以立即交付一个结果，不必先计算结果，因为它已经在队列中准备好了。

put(...) 和 take( ) 方法在收到中断时会抛出异常。在这种情况下，程序中没有特殊处理，只有 IllegalStateException。当制造者线程死亡，迭代器永远等待该元素时会产生问题。poll(long timeout, TimeUnit unit) 提供了一种替代方法，允许指定最大等待时间。在我们的例子中，这可能意味着：如果一秒后队列中没有元素——计算永远不会花费那么长时间——则我们可以认为线程已经死亡。

第 4 章
Java Stream API

Stream API 可以对数据进行分步处理。在一个源发出数据后，不同的步骤紧随其后，过滤和转换数据，并将其简化为一个结果。

Stream——中文可译作流，尽管该术语含糊不清并且可能与输入/输出流混淆——是 Java 8 的一项重大创新，并借鉴了 Java SE 库中的其他创新，如预定义函数式接口或 Optional。它与 Lambda 表达式和方法引用一起使用，会产生紧凑的代码和一种以声明形式配置处理步骤的全新方式。

该任务块中的第一个任务使用了我们之前在关于类库的章节中遇到的超级英雄。所有主要的终端和中间操作都用于这个超级英雄集合。接下来是不同的任务，其解决方案显示了 Stream API 提供的可能性有多棒。

本章使用的数据类型如下：

- java.util.stream.Stream (https://docs.oracle.com/en/java/javase/11/docs/api/java.base/java/util/stream/Stream.html)
- java.util.stream.IntStream (https://docs.oracle.com/en/java/javase/11/docs/api/java.base/java/util/stream/IntStream.html)
- java.util.stream.Collectors (https://docs.oracle.com/en/java/javase/11/docs/api/java.base/java/util/stream/Collectors.html)
- java.util.IntSummaryStatistics (https://docs.oracle.com/en/java/javase/11/docs/api/java.base/java/util/IntSummaryStatistics.html)
- java.util.DoubleSummaryStatistics (https://docs.oracle.com/en/java/javase/11/docs/api/java.base/java/util/DoubleSummaryStatistics.html)
- java.util.regex.Pattern (https://docs.oracle.com/en/java/javase/11/docs/api/java.base/java/util/regex/Pattern.html)

4.1 常规流及其终端和中间操作

每个流都有两个强制性步骤和任意数量的可选步骤：

1. 来自数据源的流的创建。
2. 可选处理步骤，称为中间操作。
3. 最终操作，称为终端操作。

4.1.1 超级英雄史诗：认识 Stream API ★

《Ciao Ciao 船长征服 Java：喂养大脑更好的 Java 技能练习卷 I》第 10 章 "Java 库中的特殊类型"介绍了类 Heroes 和超级英雄，本任务以此为基础。

创建 Stream：

始终为以下任务用超级英雄构建一个新 Stream，然后根据以下模式应用终端和中间操作：

```
Heroes.ALL.stream().intermediate1(…).intermediate2(…).terminal()
```

终端操作：

1. 在屏幕上以 CSV 格式输出有关超级英雄的所有信息。
2. 询问是否所有超级英雄都是在 1900 年以后被引入的。
3. 询问在 1950 年（含）以后有没有引入过女超级英雄。
4. 哪个超级英雄最先出现？
5. 哪个超级英雄的出现年份最接近 1960 年？在 Stream 上只能使用一个终端操作。
6. 创建一个 StringBuilder，其中包含用逗号分隔的所有年份数字。使用单终端 Stream 方法获得结果，没有中间操作。字符串中年份数字的顺序无关紧要。
7. 将男、女超级英雄分成两组。结果应该是类型 Map<Sex, List<Hero>>。
8. 以 1970 年为界限，为在该界限之前和之后出现的超级英雄制作两个分区。结果应该是类型 Map<Boolean, List<Hero>>。

中间（非终端）操作：

1. 共有多少个女超级英雄？
2. 按发布日期对所有超级英雄进行排序，然后输出所有超级英雄。
3. 执行以下步骤：
   - 用所有女超级英雄的名字创建一个以逗号分隔的字符串。
   - Hero 中没有 Setter，因为 Hero 是不可变的。但是，使用构造函数我们可以构建新的超级英雄。将超级英雄转换为匿名超级英雄列表，其中括号中的真实姓名与括号本身一起被删除。
   - 创建一个 int[] 包含所有引入超级英雄的年份——没有重复的条目。
4. 遍历 UNIVERSES 而不是 ALL，输出所有超级英雄的姓名。

4.1.2 测试：双倍输出 ★

如果以下 3 行在 main(...) 方法中并且程序启动，输出将是什么？

```
Stream<Integer> numbers = Stream.of( 1, 2, 3, 4, 5 );
numbers.peek( System.out::println );
numbers.forEach( System.out::println );
```

4.1.3 从列表中找出心爱的船长 ★

年底，船员们投票决定哪位船长候选人应该在未来为他们铺设致富之路。获胜者是被提名最多的人。

任务：
- 给定一个带有姓名的字符串数组。姓名被提及的频率如何？不区分大小写。
- 很多人把 CiaoCiao 船长称为 CiaoCiao，这应该视作 Captain CiaoCiao 的同义词。

举例：
{ "Anne", "Captain CiaoCiao", "Balico", "Charles", "Anne", "CiaoCiao", "CiaoCiao", "Drake", "Anne", "Balico", "CiaoCiao" } → {charles=1, anne=3, drake=1, ciaociao=4, balico=2}

4.1.4 框住图片（Java 11）★

CiaoCiao 船长被评为最佳船长，其喜悦之情溢于言表。他希望将自己的照片装裱起来。

给出一个多行字符串，例如：

```
                 _____
           _.-':::::::'.
          \:::::::::::::'.-._
           \:::''    `::::`-.`.
            \           `:::::`.\
             \             `-:::::`:
              _____          `:::`.
               .|_.-'__`._        `:::\
              ,'`|:::| )/`.        \:::
             /. -.`--'  : /.\       ::|
             `-,-'  _,'/| \|\\      |:|
             ,'`::.   |/>`;'\       |:|
             (_\ \:.:.:`((_));`. ;:|
             \.:\ ::_:_:`-',' `-:|
              `:\\|   SSt:
                )`__...---'
```

将它放在一个框架中，例如：

```
+-----------------------------+
|                             |
|         _____              |
|   _.-':::::::'.             |
|  \:::::::::::::'.-._        |
|   \:::''    `::::`-.`.      |
|    \           `:::::`.\    |
|     \             `-:::::`: |
|      _____          `:::`.|
|       .|_.-'__`._        `:::\
|      ,'`|:::| )/`.        \:::|
|     /. -.`--'  : /.\       ::| |
|     `-,-'  _,'/| \|\\      |:| |
```

任务：

编写方法 frame(String)，框住一个多行字符串。使用 String 方法 lines( ) 和 repeat( )。这两种方法自 Java 11 起可获取。如果你想使用 Java 8 实现该任务，则必须换个思路。

▶ 水平线由 "-" 组成。

- 垂直线由"|"组成。
- 角落使用加号"+"。
- 框架左、右的空间为 2 空格。
- 内部上、下空间为 1 空行。
- 换行符是 \n，但在程序中可以轻易修改。

4.1.5 看和说★★

CiaoCiao 船长正在做白日梦，并在一张纸上写字：

1

他看到 1，并对自己说："哦，1 个 1！"他把这写了下来：

1 1

现在他看到了 2 个 1，可以像这样读出来：

2 1

"啊！那是 1 个 2 和 1 个 1！"他把它写了下来：

1 2 1 1

他又看了看数字，说：

1 1 1 2 2 1

现在 1 甚至出现了 3 次：

3 1 2 2 1 1

CiaoCiao 船长认为，这些数字正在迅速变大。他很好奇，如果再这样下去，是否只出现数字 1，2 和 3。

任务：

- 使用 Stream.iterate(...) 生成无限的看和说数字 Stream。

- 将 Stream 限制为 20 个元素。
- 在控制台输出数字；它们也可以用 "111221" 这样的字符串紧凑输出。

> **小提示：**
> 该任务可以通过带有反向引用的正则表达式巧妙地完成。但是，此解决方案要求很高，任何想选择该方案的读者都可以在 https://regular-expressions.mobi/backref.html 找到更多详细信息。

> **提示：**
> 该任务涉及 Look-and-Say-Sequenz（看和说顺序），https://oeis.org/A005150 用许多参考资料对其进行了更详细的解释。

4.1.6　删除包含稀土金属的重复岛屿 (Java 9) ★★★

稀土金属业务对 Bonny Brain 特别有吸引力。她的船员编制了一份清单，列出了在哪些岛屿上发现了哪些稀土金属。结果显示在一个如下所示的文本文件中：

```
Balancar
Erbium
Benecia
Yttrium
Luria
Thulium
Kelva
Neodym
Mudd
Europium
Tamaal
Erbium
Varala
Gadolinium
Luria
Thulium
```

一行内容为岛屿，下一行为稀土金属。但是，不同的船员可能在文本文件中输入相同的对。在示例中是 Luria 和 Thulium 对重复。

任务：
- 编写一个程序，从文本中删除所有重复的行对。
- 程序必须足够灵活，输入可以来自 String、File、InputStream 或 Path。
- 行总是由 \n 分隔。最后一行也以 \n 结尾。

> **小提示：**
> Pattern、Scanner 和 MatchResult 类型以及 Scanner 方法 findAll(...) 和其他 Stream 方法有助于解决问题。findAll(...) 是 Java 9 版本的新功能。

4.1.7 船帆在哪里？ ★★

Bonny Brain 需要一个新的高性能船帆。负责材料管理的文员准备了一份清单，上面有合适的布料制造商的坐标：

```
Point.Double[] targets = { // Latitude, Longitude
    new Point.Double( 44.7226698, 1.6716612 ),
    new Point.Double( 50.4677807, -1.5833018 ),
    new Point.Double( 44.7226698, 1.6716612 )
};
```

任务：
- 在列表中，有些坐标出现了两次，它们可以忽略。
- 在最后，应该有一个 Map<Point.Double, Integer>，其中有坐标以及 Bonny Brain 到当前位置的距离（千米）（40.2390577, 3.7138939）。

输出可以是这样的：

```
{Point2D.Double[50.4677807, -1.5833018]=1209, Point2D.Double[44.7226698, 1.6716612]=525}
```

以千米为单位的距离用半正矢公式计算，像这样：

```
private static int distance( double lat1, double lng1,
```

```
    double lat2, double lng2 ) {
double earthRadius = 6371; // km
double dLat = Math.toRadians( lat2 - lat1 );
double dLng = Math.toRadians( lng2 - lng1 );
double a = Math.sin( dLat / 2 ) * Math.sin( dLat / 2 ) +
  Math.cos( Math.toRadians( lat1 ) ) * Math.cos( Math.toRadians( lat2 ) ) *
  Math.sin( dLng / 2 ) * Math.sin( dLng / 2 );
double d = 2 * Math.atan2( Math.sqrt( a ), Math.sqrt( 1 - a ) );
return (int) (earthRadius * d);
}
```

4.1.8 购买最受欢迎的装甲车★★★

CiaoCiao 船长需要扩充他的舰队，他问船员他们推荐什么装甲车。CiaoCiao 船长获取了以下类型的模型名称数组：

```
String[] cars = {
  "Gurkha RPV", "Mercedes-Benz G 63 AMG", "BMW 750", "Toyota Land Cruiser",
  "Mercedes-Benz G 63 AMG", "Volkswagen T5", "BMW 750", "Gurkha RPV",
  "Dartz Prombron", "Marauder", "Gurkha RPV" };
```

任务：
- 编写程序，处理一个模型名称数组并最终生成 Map<String, Long>，它将模型名称与出现次数关联。这部分任务可以通过 Stream API 很好地完成。
- 不应该有只提名一次的模型；只有具有两个或更多条目的模型才能出现在数据结构中。在这部分任务中，我们可以放弃 Stream API，而使用另一个变体。

输出可能如下所示：

```
{Mercedes-Benz G 63 AMG=2, BMW 750=2, Gurkha RPV=3}
```

修改查询，使所有模型都在一个 Map 中，但如果提名少于两次，则名称与 false 关联。输出可能如下所示：

```
{Marauder=false, Dartz Prombron=false, Mercedes-Benz G 63 AMG=true, Toyota Land Cruiser=false, Volkswagen T5=false, BMW 750=true, Gurkha RPV=true}
```

4.2 原始流

除了对象流之外，Java 标准库还为原始数据类型提供了三种特殊流：IntStream、LongStream 和 DoubleStream。许多方法是相似的，重要的区别是范围（英语：range）和特殊精简，例如求和或平均。

4.2.1 识别数组中的非数字★

Java 对 double 浮点类型支持三种特殊值：Double.NaN，Double.NEGATIVE_INFINITY 和 Double.POSITIVE_INFINITY；Float 中有 float 对应的常量。对于数学运算，需要检查结果是否有效而不是 NaN。除非操作数是 NaN，否则算术运算（如加法、减法、乘法、除法）无法得到 NaN，但 Math 类中的各种方法返回 NaN 作为无效输入的结果。例如，在方法 log(double a) 或 sqrt(double a) 中，如果参数 a 小于零，则结果为 NaN。

任务：

▶ 编写方法 containsNan(double[])，如果数组包含 NaN，则返回 true，否则返回 false。
▶ 在方法的主体中，一个表达式就足够了。

举例：
```
double[] numbers1 = { Math.sqrt( 2 ), Math.sqrt( 4 ) };
System.out.println( containsNan( numbers1 ) );          // false

double[] numbers2 = { Math.sqrt( 2 ), Math.sqrt( -4 ) };
System.out.println( containsNan( numbers2 ) );          // true
```

4.2.2 生成数十年★

十年是指十年的时间，不管它在何时开始和结束。在通常情况下，十年是按其共同的十位数分组的。20 世纪 90 年代从 1990 年 1 月 1 日开始，到 1999 年 12 月 31 日结束。这种解释被称为 0-9 十年。还有一种是 1-0 十年，即在第 1 位数上以 1 开始计算年代。根据这种解释，20 世纪 90 年代从 1991 年 1 月 1 日开始，到 2000 年 12 月 31 日结束。

任务：

▶ 编写方法 int[] decades(int start, int end)，以数组的形式返回从开始年份到结束年份的所有十年。

- 使用 0-9 十年。

举例：
- Arrays.toString(decades(1890, 1920)) → [1890, 1900, 1910, 1920];
- Arrays.toString(decades(0, 10)) → [0, 10];
- Arrays.toString(decades(10, 10)) → [10];
- Arrays.toString(decades(10, -10)) → [ ]。

4.2.3　通过 Stream 创建具有恒定内容的数组★

任务：
编写方法

```
fillNewArray(int size, int value)
```

举例：
```
Arrays.toString( fillNewArray( 3, -1 ) ) → [-1, -1, -1]
```

4.2.4　绘制金字塔（Java 11）★

任务：
- 巧妙组合 range(...)，mapToObj(...) 和 forEach(...) 生成以下输出：

```
    /\
   /\/\
  /\/\/\
 /\/\/\/\
/\/\/\/\/\
```

金字塔高 5 行。
- 尝试用一条指令完成任务——从金字塔构建到控制台输出。

4.2.5　查找字符串的字母频率★

压缩的前提是尽可能短地表示频繁出现的序列。例如，如果在一个文件中出现了 0 0 0 0 1 1 1，那么就会存储以下内容：先是 4 个 0，然后是 3 个 1。压缩算法试图用极少的比特来表达关于数字 0 和 1 的信息。其优点是如果知道一个符号或一个序列总共出现的频率，就能估计这个序列的压缩是否值得。一个循环可以事先运行输入，并计算频率。

任务：

▶ 输入是一个字符串。使用巧妙的 Stream 连接，生成一个新的字符串，包含源字符串的每个字母，后跟该字母在给定字符串中的频率。
▶ 在结果字符串中，字母和频率的配对应以斜线分开。
▶ 性能并不是最重要的。

举例：

▶ "eclectic" → "e2/c3/l1/e2/c3/t1/i1/c3";
▶ "cccc" → c4/c4/c4/c4;
▶ "" → ""。

4.2.6 从1到0，从10到9 ★★

Bonny Brain 想购买一艘新船，并派伊莱恩到港口评估船只。伊莱恩在一张纸上写下她的评分（从 1 到 10），如下所示：

102341024

Bonny 得到了数字序列，但对排列和数字不满意。首先，数字应该用逗号分隔；其次，它们应该从 0 开始，而不是从 1 开始。

任务：

▶ 编写方法 String decrementNumbers(Reader)，从输入源读取一串数字并将其转换为逗号分隔的字符串；所有数字都应减 1。结果中不包含不是数字的内容。

举例：

▶ 102341024 → "9, 1, 2, 3, 9, 1, 3";
▶ -1 → "0";
▶ abc123xyz456 → "0, 1, 2, 3, 4, 5"。

4.2.7 合并两个 int 数组 ★★

任务：

▶ 寻找方法合并两个 int 数组。
▶ 应有两种重载方法：
```
static int[] join( int[] numbers1, int[] numbers2) und
static int[] join( int[] numbers1, int[] numbers2, long maxSize).
```

使用可选的第三个参数，可以减少结果的最大元素数。

举例：
```
int[] numbers1 = { 7, 12 };
int[] numbers2 = { 51, 56, 0, 2 };
int[] result1 = join( numbers1, numbers2 );
int[] result2 = join( numbers1, numbers2, 3 );
System.out.println( Arrays.toString( result1 ) ); // [7, 12, 51, 56, 0, 2]
System.out.println( Arrays.toString( result2 ) ); // [7, 12, 51]
```

4.2.8 确定获胜组合★★

Bonny Brain 计划下一次聚会，并准备了一场抛环游戏。首先，她设置了不同的对象，例如：

▨▨

然后，她给玩家两个圆环，让他们投掷。如果圆环越过一个物体，则算作胜利。有多少种获胜方式？机会有多大？取两个对象▨▨，玩家可以命中▨或▨，或同时命中▨和▨；没有命中就不算获胜。

任务：
- 给出一个包含来自基本多语言平面 (BMP) 的任意字符的字符串，即从 U+0000 到 U+D7FF 和从 U+E000 到 U+FFFF。
- 列出玩家可以获胜的所有可能性。

举例：
- ▨▨ → [▨, ▨, ▨▨]，但不是 [▨, ▨, ▨▨, ▨▨]！
- ▣◻▲ → [▣◻▲, ▲, ◻, ▣, ◻▲, ▣▲, ▣◻]。
- MOON → [OO, MN, MO, MOON, MOO, MON, M, N, OON, ON, O]。

4.3 统计数据

IntStream，LongStream 和 DoubleStream 流具有终止方法，例如 average( )，count( )，max( )，min( ) 和 sum( )。然而，如果不是对一个统计信息有兴趣，而是几个，则各种信息可以被收集在 IntSummaryStatistics，LongSummaryStatistics 或 DoubleSummaryStatistics 中。

4.3.1 最快和最慢的桨手 ★

Bonny Brain 在派对岛举办年度划船比赛"毒蛇划船公开赛"。最后，要给出最长时间、最短时间和平均时间。桨手们的成绩体现为以下几个数据类型：

```
class Result {
  String name;
  double time;

  Result( String name, double time ) {
    this.name = name;
    this.time = time;
  }
}
```

任务：

▶ 创建一个 Result 对象的 Stream。用选定的值预先赋值一些 Result 对象进行测试。

▶ 最后输出一个小的时间统计。

举例：
Stream 如下：

```
Stream<Result> stream =
    Stream.of( new Result( "Bareil Antos", 124.123 ),
      new Result( "Kimara Cretak", 434.22 ),
      new Result( "Keyla Detmer", 321.34 ), new Result( "Amanda Grayson",
        143.99 ),
      new Result( "Mora Pol", 122.22 ), new Result( "Gen Rhys", 377.23 ) );
```

输出如下：

```
count: 6
min: 122,22
max: 434,22
average: 253,85
```

4.3.2 计算中位数 ★★

XXXSummaryStatistics 类型用 getAverage( ) 返回算术平均值。算术平均值是由给定值的总和除以数值的数量计算出来的。还有一些其他的平均数，如几何平均数或调和平均数。

平均值在统计学中经常使用，但它们有一个问题，即更容易出现离群值。统计学经常用中位数来工作。中位数是中心值，即在排序的列表中处于"中间"的值。过小或过大的数字都处于边缘，属于离群值，不包括在中位数内。

如果值的数量是奇数，则存在自然均值。

举例：
- 在列表 9,11,11,11,12 中中位数为 11。如果值的数量是偶数，则中位数可以被定义为中间两个数的算术平均值。
- 在列表 10,10,12,12 中，中位数是值 10 和 12 的算术平均值，即 11。

任务：
- 给定带有测量值的 double[ ]。编写方法 double median (double... values)，计算偶数和奇数数量的数组的中位数。
- 使用 DoubleStream 解决问题，并考虑 limit(...) 和 skip(...) 是否有所帮助。

4.3.3 统计温度并绘制图表 ★★★

Bonny Brain 擅长算数，但她更喜欢图表。以图形形式理解数据比以文本形式理解数据容易得多。

她收到一张包含温度数据的表格，希望一眼就能看出什么时候最热，什么时候适合家庭度假。

任务：
我们寻找的是可以处理和显示温度的程序。进一步来说：
- 生成一个随机数列表，最好的情况是这些随机数遵循一年的温度曲线规律，例如以从 0 到 π 的正弦曲线规律。
- 生成几年的随机温度值，并将这些年份与值一起存储在关联映射中。使用 Year 数据类型作为按年份排序的 Map 的键。额外奖励：天数对应一年中的实际天数，即 365 或 366。
- 将包含所有年份温度的 ASCII 表写入控制台。
- 报告一年中的最高和最低年温度。
- 报告一年中一个月的最高、最低和平均温度。

▶ 生成一个文件，其中汇总了一年十二个月的每月平均温度并进行了可视化。以下面的 HTML 文档为基础，相应地填充 data 数组：

```
<!DOCTYPE html>
<html lang="de">
<body>
<canvas id="chart" width="500" height="200"></canvas>
<script src="https://cdnjs.cloudflare.com/ajax/libs/Chart.js/2.9.3/Chart.bundle.min.js"></script>
<script>
const cfg = {
    type: 'bar',
    data: {
        labels: ["Jan", "Feb", "Mrz", "Apr", "Mai", "Jun", "Jul", "Aug",
                "Sep", "Okt", "Nov", "Dez"],
        datasets: [ {
            label: "Durchschnittswerte", fill: false,
            data: [11.9,17.0,21.3,25.1,27.8,29.1,29.2,27.9,25.6,21.6,
                   17.5,12.5],
        } ]
    },
    options: {
        responsive: true,
        title: { display:true, text:'Temperaturverlauf' },
        tooltips: { mode: 'index', intersect: false },
        hover: { mode: 'nearest', intersect: true },
        scales: {
            xAxes: [ { display: true, scaleLabel: { display: true,
                        labelString: 'Monat' } } ],
            yAxes: [ { display: true, scaleLabel: { display: true,
                        labelString: 'Temperatur' } } ]
        }
    }
};
window.onload = () =>
    new Chart(document.getElementById("chart").getContext("2d"), cfg);
```

```
</script>
</body>
</html>
```

4.4 可供参考的解决方案

任务 4.1.1：超级英雄史诗：认识 Stream API

下面引入一个辅助变量，作为 Heroes.ALL 的缩写：

```
List<Hero> heroes = Heroes.ALL;
```

终端操作 forEach(...)

```
Consumer<Hero> csvPrinter =
    hero -> System.out.printf( "%s,%s,%s%n", hero.name, hero.sex,
    hero.yearFirstAppearance );
heroes.stream().forEach( csvPrinter );
```
列表 4.1 com/tutego/exercise/stream/LambdaHeroes.java

forEach(...) 方法可用于遍历 Stream 的所有元素：

```
void forEach(Consumer<? super T> action)
```

它需要一个 Consumer 类型的消费者。forEach(Consumer) 方法为每个元素调用传递的 Consumer 的 apply(...) 方法，并以这种方式将 Stream 中的元素传输给 Consumer。对于 Consumer，我们声明一个实现控制台输出的映射，从 Hero 中提取三个组件。我们不必为了运行而提取 Stream，Iterable 提供了 forEach(...)。

终端操作 allMatch(...)

```
Predicate<Hero> isAppearanceAfter1900 =
    hero -> hero.yearFirstAppearance >= 1900;
System.out.println( heroes.stream().allMatch( isAppearanceAfter1900 ) );
```
列表 4.2 com/tutego/exercise/stream/LambdaHeroes.java

Stream 提供了三种方法来确定 Stream 中的所有或某些元素是否具有属性：allMatch(...)、anyMatch(...) 和 noneMatch(...)。所有方法都被赋予一个用于测试属性的 Predicate。如果我们想知道是否所有超级英雄都是在 1900 年之后被引入的，则可以使用 allMatch(...) 方法。allMatch(...) 方法遍历 Stream 的所有元素并调用 Predicate 方法 test(...)。如果此测试始终返回 true，则 Stream 中的所有元素都符合正确的标准，并且总体答案为 true。如果其中一项测试返回 false，则最终结果已经确定：false。

终端操作 anyMatch(...)

```
Predicate<Hero> isFemale = hero -> hero.sex == Sex.FEMALE;
Predicate<Hero> isAppearanceAfter1950 =
    hero -> hero.yearFirstAppearance >= 1950;
System.out.println(
    heroes.stream().anyMatch( isFemale.and(isAppearanceAfter1950 ) ) );
```

列表 4.3 com/tutego/exercise/stream/LambdaHeroes.java

但是，Predicate 接口有一些 default 和 static 方法，包括 and(Predicate)，or(Predicate) 和 negate( )。如果要同时应用两个条件，则可以使用 and(...) 方法将两个谓词连接起来，从而组合成一个更大的谓词。如果我们想测试某个超级英雄是否是女性并且是在 1950 年之后被引入的，则可以首先建立两个单独的谓词，然后将它们连接起来。这种方法很有意义：首先，谓词更小，更容易测试；其次，它们易于再次使用。

终端操作 min(...):

```
Comparator<Hero> firstAppearanceComparator =
    Comparator.comparingInt( h -> h.yearFirstAppearance );
System.out.println( heroes.stream().min( firstAppearanceComparator ) );
```

列表 4.4 com/tutego/exercise/stream/LambdaHeroes.java

Stream 提供 min(...) 和 max(...) 方法，并且可以使用排序标准确定最大和最小元素：

```
Optional<T> min(Comparator<? super T> comparator)
Optional<T> max(Comparator<? super T> comparator)
```

两种方法的返回都是 Optional，因为 Stream 可能是空的。min(...) 和 max(...) 方法需要一个 Comparator。在询问最先被引入的超级英雄时，标准是 yearFirstAppearance。Comparator.comparingInt(...) 有助于快速构建 Comparator，年份越小，对象越小。min(...) 方法返回最早被引入的超级英雄。

终端操作 reduce(...):

```
System.out.println( heroes.stream().reduce( ( hero1, hero2 ) -> {
```

```
        int diff1 = Math.abs( 1960 - hero1.yearFirstAppearance );
        int diff2 = Math.abs( 1960 - hero2.yearFirstAppearance );
        return (diff1 < diff2) ? hero1 : hero2;
} ) );
```

列表 4.5 com/tutego/exercise/stream/LambdaHeroes.java

关于哪个超级英雄在 1960 年前后首先出现的问题，我们求助于 reduce(...) 方法，它恰恰倾向于那些出现时间更接近 1960 年的超级英雄。reduce(...) 声明如下：

```
Optional<T> reduce(BinaryOperator<T> accumulator)
```

必须传递 BinaryOperator。这意味着两个元素作为参数进来，一个元素出来。选择的实现首先计算从第一个和第二个超级英雄到 1960 年的距离，然后返回出现时间更接近 1960 年的超级英雄。如果两者与 1960 年的距离相等，则原则上可以选择两个超级英雄之一。

终端操作 collect(Supplier, BiConsumer, BiConsumer)：
```
StringBuilder collectedYears =
    heroes.stream().collect(
        StringBuilder::new,
        ( sb, hero ) -> sb.append( sb.isEmpty() ? "" : "," )
            .append( hero.yearFirstAppearance ),
        ( sb1, sb2 ) -> sb1.append( sb2.isEmpty() ? "" : "," + sb2 ) );
System.out.println( collectedYears );
```
列表 4.6 com/tutego/exercise/stream/LambdaHeroes.java

如果最后要创建一个所有值的 StringBuilder，则我们不会使用 reduce(...) 方法，而是使用 collect(...) 方法。虽然 reduce(...) 总是将两个值归约为一个值，但 collect(...) 会查看所有元素并将它们转换为另一种表示形式。该方法重载，但我们感兴趣的变体如下：

```
R collect(Supplier<R> supplier,
        BiConsumer<R, ? super T> accumulator,
        BiConsumer<R, R> combiner)
```

第一个参数是结果的生产者。因为在我们的例子中结果是一个 StringBuilder，

StringBuilder::new 构造函数引用创建了这个 Supplier。第二个参数是 BiConsumer，将超级英雄出现的年份放入 StringBuilder。在必要时作特殊处理，即在元素之间添加逗号；当 StringBuilder 包含元素且不为空时，分隔符才会出现在元素之间。collect(...) 的最后一个 BiConsumer 可能通过并发处理创建的各种 StringBuilder 组合成一个 StringBuilder。在我们的例子中这不是必需的，但我们也想实现这个功能。

终端操作 collect(Collector) with groupingBy(...)：

```
Map<Sex, List<Hero>> sexListMap =
    heroes.stream().collect( Collectors.groupingBy( hero -> hero.sex ) );
System.out.println( sexListMap );
```

列表 4.7 com/tutego/exercise/stream/LambdaHeroes.java

第二个 collect(...) 方法参数化如下：

```
<R, A> R collect(Collector<? super T, A, R> collector)
```

Collector 被传递。Collectors 类为预定义的 Collector 实现声明了大量静态方法。其中包括 toList( )，toSet( )。Collector 很实用，可用于对 Stream 元素进行分组。Collectors.groupingBy(...) 如下：

```
<T, K> Collector<T, ?, Map<K, List<T>>> groupingBy(
  Function<? super T, ? extends K> classifier)
```

泛型类型信息有点晦涩，但实际上方法很简单：Function 的任务是提取生产 Map 的键。Stream 中具有相同键的所有元素都作为列表与 Map 中的键关联。

如果因此需要具有性别的 Map，则函数是 hero –> hero.sex，并创建一个 Map<Sex, List<Hero>>，以便列出男性或女性超级英雄的列表。

终端操作 collect(Collector) with partitioningBy( )：

```
Predicate<Hero> isAppearanceAfter1970 =
  hero -> hero.yearFirstAppearance >= 1970;
Map<Boolean, List<Hero>> beforeAndAfter1970Partition = heroes.stream()
    .collect( Collectors.partitioningBy( isAppearanceAfter1970 ) );
System.out.println( beforeAndAfter1970Partition );
```

列表 4.8 com/tutego/exercise/stream/LambdaHeroes.java

groupingBy(...) 的结果始终是具有任意数量键的 Map。

如果产生的集合只知道两个不同的部分，则可以使用 partitioningBy(...) 方法作为替代方法：

```
Collector<T, ?, Map<Boolean, List<T>>> partitioningBy(
    Predicate<? super T> predicate)
```

Predicate 传递给测试，通过此测试的元素作为键放入 Map 的 Boolean.TRUE，其他元素放在 Boolean.FALSE 中。

这就回答了 1970 年前后出现了哪些超级英雄的问题。

中间操作 filter(...):

```
System.out.println( Heroes.ALL.stream()
    .filter( hero -> hero.sex == Sex.FEMALE )
    .count() );
```

列表 4.9 com/tutego/exercise/stream/LambdaHeroes.java

可能最重要的中间 Stream 方法是 filter(...):

```
Stream<T> filter(Predicate<? super T> predicate);
```

满足谓词的所有元素都保留在 Stream 中。如果问题询问女超级英雄，则我们编写一个谓词来提取超级英雄的性别并测试 Sex.FEMALE。创建一个新 Stream，但经过过滤后，该 Stream 可能包含较少的元素。count() 方法返回元素的数量。

中间操作 sorted(...):

```
Heroes.ALL.stream()
    .sorted( Comparator.comparingInt( hero -> hero.yearFirstAppearance ) )
    .forEach( System.out::println );
```

列表 4.10 com/tutego/exercise/stream/LambdaHeroes.java

Stream 类的一些方法是有状态限制的。这些操作在被评估之前不能尽情地等待（它们不是懒惰），而是必须读入 Stream 的所有元素，以便可以实现诸如排序或删除重复元素之类的操作。如果要根据超级英雄被引入的时间对超级英雄流进行排序，则我们将首先使用 Comparator.comparingInt(...) 再次创建一个 Comparator，并将其传递给 sorted(...) 方法：

```
Stream<T> sorted(Comparator<? super T> comparator)
```

使用 forEach(...)：我们使用排序后的 Stream 并将所有超级英雄输出到屏幕上。

中间操作 map(...) with Collectors.joining(...)：

```
String femaleNames = Heroes.ALL.stream()
    .filter( hero -> hero.sex == Sex.FEMALE )
    .map( hero -> hero.name )
    .collect( Collectors.joining( ", " ) );
System.out.println( femaleNames );
```

列表 4.11 com/tutego/exercise/stream/LambdaHeroes.java

除了 filter(...)，map(...) 方法可能是 Stream 接口中第二重要的：

```
<R> Stream<R> map(Function<? super T, ? extends R> mapper)
```

map(...) 方法将 Function 应用于 Stream 中的每个元素，并创建一个新 Stream，它可能是一种新类型。因此，我们也可以使用 map(...) 方法提取所有超级英雄的姓名。在所需的过滤操作之后，姓名被提取，然后使用特殊的 Collector 将它们绑定在一起，形成一个大 String，其中姓名用逗号分隔。

中间操作 map(...) with Collectors.toList( )：

```
Function<Hero, Hero> nameAnonymizer = hero ->
    new Hero( hero.name.replaceAll( "\\s*\\(.*\\)$", "" ),
              hero.sex, hero.yearFirstAppearance );
System.out.println( Heroes.ALL.stream().map( nameAnonymizer )
    .collect(Collectors.toList() ) );
```

列表 4.12 com/tutego/exercise/stream/LambdaHeroes.java

在任务中，应删除圆括号中的条目。我们也使用 map(...) 方法和一个交换超级英雄的特殊函数来做到这一点。该函数获得了一个带有姓名的超级英雄，并返回一个删除括号中所有内容的新的超级英雄。原则上，这种处理链可以修改对象并返回修改后的对象，但创建具有所需更改的新对象更为简洁。由于 Hero 对象是不可变的，所以还需要构建新的 Hero 对象。我们提取名称并使用 replace All(...) 将括号中的所有内容以及它之前的任何空格序列替换为空字符串，即删除这部分。构造函数传递这个修改后的姓名，留下性别和年份。Collectors.toList( ) 将新创建的 Stream<Hero> 传输到一个列表中。

除了 Function<Hero, Hero>，也可以使用 UnaryOperator<Hero>。

中间操作 mapToInt(...)：

```
int[] years = Heroes.ALL.stream()
    .mapToInt( hero -> hero.yearFirstAppearance )
    .distinct()
    .toArray();
System.out.println( Arrays.toString( years ) );
```
列表 4.13 com/tutego/exercise/stream/LambdaHeroes.java

除了 map(...) 方法之外，还有一些返回特殊原始流的方法：

▶ IntStream mapToInt(ToIntFunction<? super T> mapper);
▶ LongStream mapToLong(ToLongFunction<? super T> mapper);
▶ DoubleStream mapToDouble(ToDoubleFunction<? super T> mapper)。

这三个特殊的流提供了 toArray( ) 方法，该方法以原始数组而不是对象数组结束。这是在数组中收集所有年份数字的好方法。distinct( ) 删除重复元素。

中间操作 flatMap(...)：

```
Heroes.UNIVERSES.stream()
    .flatMap( Heroes.Universe::heroes )
    .map( hero -> hero.name )
    .forEach( System.out::println );
```
列表 4.14 com/tutego/exercise/stream/LambdaHeroes.java

map(...) 方法的传递函数进行直接映射。这不会创建更多元素，只是交换元素。flatMap(...) 方法则不同：

```
<R> Stream<R> flatMap(Function<? super T, ? extends
    Stream<? extends R>> mapper)
```

map(...) 传递的是 Function<? super T, ? extends R>，而 flatMap(...) 传递的是 Function<? super T, ? extends Stream<? extends R>>；也就是说，在 flatMap(...) 中，函数必须为每个元素返回一个 Stream。flatMap(...) 运行这些子 Stream，直接把元素放

在结果 Stream 中。换句话说，flatMap(...) 返回一个带有子元素的 Stream<...>，而 map(...) 则会返回一个 Stream<Stream<...>>。flatMap(...) 提供了一种方法，使产生的 Stream 增长和变大。

Heroes.UNIVERSES 是一个 List<Universe>。如果我们请求一个 Stream，则结果是 Stream<Universe>。一个 Universe 有 String name( ) 和 Stream<Hero> heroes( ) 方法。如果我们对两个宇宙的所有 Hero 对象感兴趣，则 flatMap(Heroes.Universe::heroes) 会有帮助，因为它返回一个 Stream<Hero>。作为比较，map(Heroes.Universe::heroes) 返回一个 Stream<Stream<Hero>>，我们不能用它做任何事情。

由于在任务中只有姓名是相关的，所以额外的 map(...) 会导致一个 Stream<String>，forEach(...) 输出姓名。

测试 4.1.2：双倍输出

运行该程序会导致异常：

```
Exception in thread "main" java.lang.IllegalStateException: stream has already been operated upon or closed
```

Stream 对象的中间操作返回新的 Stream 对象，我们必须以级联方式调用进一步的方法。如果意图是两个流都通过 peek(...) 和 forEach(...) 重新开始，则还必须重建 Stream。否则，如果 peek(...) 已经启动了 Stream numbers，则 forEach(...) 无法重新开始。

任务 4.1.3：从列表中找出心爱的船长

```
String[] names = {
    "Anne", "Captain CiaoCiao", "Balico", "Charles", "Anne", "CiaoCiao",
    "CiaoCiao", "Drake", "Anne", "Balico", "CiaoCiao" };
Map<String, Long> nameOccurrences =
    Arrays.stream( names )
        .map( s -> "CiaoCiao".equalsIgnoreCase(s)? "Captain CiaoCiao" :s)
        .collect( Collectors.groupingBy( String::toLowerCase,
        Collectors.counting() ) );
System.out.println( nameOccurrences );
```

列表 4.15 com/tutego/exercise/stream/NameOccurrences.java

我们运行程序，输出如下：

```
{captain ciaociao=4, charles=1, anne=3, drake=1, balico=2}
```

起点是带有名称的数组 names。Arrays.stream(...) 为我们提供了一个字符串 Stream，另一种选择是 Stream.of(...)。由于我们的船长可以出现在不同的写法中，所以我们将写法标准化，并且使用 map(...) 方法，Captain CiaoCiao 总是进入 Stream，而不是 CiaoCiao。其他字符串不会被转换。

实际的聚合和计数是通过一个特殊的 Collector 进行的：

```
groupingBy( Function<? super T, ? extends K> classifier,
            Collector<? super T, A, D> downstream )
```

如果我们将 groupingBy(...)–Collector 传递给 collect(...)，则最终会得到 Map<K, D> 类型的结果。

所有 groupingBy(...) 收集都期望一个分类器作为第一个参数。这决定了生成的关联映射的键。如果我们没有在 downstream 指定第二个 Collector，则会产生一个与键关联的列表，但我们不需要它——我们只对与键关联的元素的数量感兴趣。Collectors.counting( ) 返回一个归约为 Long 而不是列表的 Collector。在内部，counting(...) 方法实现如下：

```
public static <T> Collector<T, ?, Long> counting() {
    return summingLong(e -> 1L);
}
```

列表 4.16 OpenJDK-Implementierung von counting(⋯)

counting( ) 返回一个 Collector<T, ?, Long>，即一个 Collector，它接收 T 类型的元素并将它们归约为 Long。从代码中我们可以看出，这也只是一个缩写，我们也可以这样写：

```
Collectors.groupingBy( String::toLowerCase, Collectors.summingLong( e -> 1L ))
```

将一个函数传递给 summingLong(ToLongFunction)，该函数为 Stream 中的每个元素调用。通过 e -> 1L 的常量映射，summingLong(...) 只加了 1，这实际上是对计算能力的浪费。

groupingBy(Function classifier, Collector downstream) 也只是 groupingBy(classifier, HashMap::new,downstream) 的缩写。

任务 4.1.4：框住图片（Java 11）

```
private static String frame( String string ) {
  if ( string == null || string.trim().isEmpty() )
    throw new IllegalArgumentException(
      "String to frame can not be null or empty" );

  final String NEW_LINE = "\n";
  int max = string.lines().mapToInt( String::length ).max().getAsInt();
  String topBottomBorder = '+' + "-".repeat( max + 4 ) + '+' + NEW_LINE;
  String emptyRow = "| " + " ".repeat( max ) + " |" + NEW_LINE;

  return string.lines()
             .map( s -> "| " + s + " ".repeat( max - s.length() ) + " |" )
             .collect( Collectors.joining( NEW_LINE,
               topBottomBorder + emptyRow,
               NEW_LINE + emptyRow +
               topBottomBorder ) );
}
```

列表 4.17 com/tutego/exercise/stream/FramePicture.java

为了围绕 ASCII 艺术绘制框架，必须解决不同的子问题。从最长的行有多长这个问题开始，因为这决定了框架的宽度。实际答案来自单个 Stream 表达式。如果将每个字符串映射到字符串长度并从整数 Stream 中获取最大值，我们就会得到答案。

我们用这个长度做两件事：它有助于生成顶部和底部的水平边框，另外还用空格填充较短的行，以便之后所有行的长度相同。框架以加号开始，然后是减号，每边比最长的字符串多两个减号。每边都会多两个减号，因为这是到左、右边缘的（所需）内部距离。为了生成减号，我们使用 String 的 repeat(int)。顶部和底部框架不是唯一可以预先生成的 String，空白行也是如此，左、右只有一个竖线，中间是空格。我们稍后将把空白行放在顶线下方和底线上方。

准备好变量后，实际上把图片放到框架里只需一个 Stream 表达式。同样，我们使用 lines() 方法获取行的 Stream，并使用 map(...) 方法从 ASCII 图像转换每个 String：

- 将框架符号和距离设置在图像字符串的前面。
- 将空格放置在字符串之后，以便字符串始终具有相同的整体宽度。
- 之后又是一段距离和右侧的框架字符。

最后，我们收集所有行并使用利用三条信息创建的 Collector：
1. 行是怎么分隔的？换行符。
2. 整个字符串的前缀是什么？它是顶部框架，后跟一个空行。
3. 框架最后的后缀是什么？空行和框架底线。

任务 4.1.5：看和说

这个任务的美妙之处在于，它的核心部分只包括一条指令。Stream API 提供的可能性是非常吸引人的。然而，在解决方案的背后，还有一个具有非常特殊的正则表达式的 Pattern 类。

```
Pattern sameSymbolsPattern = Pattern.compile( "(.)\\1*" );
Function<MatchResult, String> lengthAndSymbol =
    match -> match.group().length() + match.group( 1 );

Stream.iterate( "1", s -> sameSymbolsPattern.matcher( s ).replaceAll(
    lengthAndSymbol ) )
    .limit( 20 )
    .forEach( System.out::println );
```

列表 4.18 com/tutego/exercise/stream/LookAndSay.java

以字符串 111221 为例。首先我们可能想到的是一个计数循环，它总是将索引前移一个位置并检查符号是否已更改。但是，我们可以在这里使用正则表达式。乍一看这很奇怪，因为不知道这里应该识别什么？从 1 到 2 再到 1 的变化？不！正则表达式可以帮助我们捕获重复的符号。我们将使用以下正则表达式编写正常重复：

.*

然而，会有一系列任意字符，但是我们必须表达同一个符号多次重复。字符本身可以是任意的。为此，可以使用正则表达式引擎实现的一个特点：反向引用。

(.)\\1*

\1 是引用第一组，即 (.) 的反向引用。在任何情况下，点匹配一个符号，反向引用匹配任何其他数量的相同符号。我们必须在这里使用 * 而不是 +，因为符号可能只出现一次，这意味着不需要反向引用。

序列的实用之处在于我们只需要查看相同符号的序列。这个符号序列包含我们需要的两条信息：某个符号出现的频率是多少？如果用正则表达式找到所有的地方，那么我们可以用字符串的长度和符号来替换发现。这正是建议解决方案所做的。

第一步，中间变量 sameSymbolsPattern 保存 Pattern。当一个程序多次需要相同的模式时，这总是很好的。第二步，我们声明一个变量 lengthAndSymbol，将 MatchResult 映射到一个 String，结果是：符号序列的长度是字符串的第一部分，重复符号是字符串的第二个字符。有趣的是，只有长度 1，2，3，因此总是有两个字符连接在一起。原则上，这两个变量都不是必需的，但它们使接下来的 Stream 更短且更易于阅读。

静态方法 Stream.iterator(...) 需要一个起始值作为第一个参数。它是 "1"。接下来是一个 UnaryOperator，一个输入和输出类型相同的 Function。使用 iterator(...) 的最后一个值调用运算符，在开头使用 "1"，然后可以从那里向上工作。总的来说，我们将 Stream 限制为 20 个元素，并在终端操作中将所有元素输出到屏幕。在每一步，字符串都会增长大约 30%，因此最好限制迭代次数。

任务 4.1.6：删除包含稀土金属的重复岛屿 (Java 9)

该任务有两个特点，所以我们不能只写

```
Arrays.stream( s.split( "\n" ) )
    .distinct().collect(Collectors.joining("\n") )
```

▶ 输入不仅可以来自 String，还可以通过 File，InputStream 或 Path 给出。
▶ 输入不是单行，而总是两行。

但是，这两个要求都不会改变构造 Stream <String> 并让 distinct( ) 删除重复条目的方法。中心问题只有：我们如何将不同的源传输到 Stream 中，以及如何将两行实现为 Stream 中的一个字符串？

非常灵活的 Scanner 类可用于将带有分隔符的字符串转换为 Stream <String>。Scanner 对象可以使用不同的输入源进行初始化，包括任务中所需的 String, File, InputStream 和 Path 类型。

```
String lines =
    "Balancar\nErbium\n" +
    "Benecia\nYttrium\n" +
    "Luria\nThulium\n" + // <-
    "Kelva\nNeodym\n" +
```

```
        "Mudd\nEuropium\n" +
        "Tamaal\nErbium\n" +
        "Varala\nGadolinium\n" +
        "Luria\nThulium\n"; // <-

// "(?m)(^.*$\n?){2}
Pattern pattern = Pattern.compile( "(^.*$\n)" + // A line
    "{2}",                                       // two lines
    Pattern.MULTILINE );
String s = new Scanner( lines )
    .findAll( pattern )
    .map( MatchResult::group )
    .distinct()
    .collect( Collectors.joining() );
System.out.println( s );
```

列表 4.19 com/tutego/exercise/stream/RemoveAllEqualPairs.java

Scanner 是一个标记器，可以以不同的方式使用：

- 读入由空格分隔的标记；
- 读取以换行符结尾的行；
- 读入基元；
- 使用确定分隔符的任意正则表达式读取标记；
- 读取匹配正则表达式的标记。

这些 nextXXX( ) 方法没有帮助，因为它们不会产生 Stream。仅从 Java 9 开始，Scanner 类才具有三个返回 Stream 的方法：

- Stream<MatchResult> findAll(String patString);
- Stream<MatchResult> findAll(Pattern pattern);
- Stream<String> tokens( )。

findAll(...) 方法有助于完成任务，因为它返回一个 Match 的结果。我们所要做的就是使用正则表达式来确定我们究竟想要捕获什么，而这正是两行。正则表达式 "(^.*$\n){2}" 由两个主要部分组成：

第 4 章　Java Stream API | 177

1. ^ 代表行首，$ 代表行尾，一行之后是换行符；
2. 这种行和换行的构造连续出现两次，由 {2} 表示。

为了使边界匹配器 ^ 和 $ 匹配本地行而不是整个输入，必须设置 Pattern.MULTILINE 标志。建议解决方案选择了这个变体，但该标志也可以直接在正则表达式中使用，我们可以编写 "(?m)(^.*$\n?){2}"。

如果此正则表达式进入 findAll(...) 方法，则会创建一个 Stream <MatchResult>（从 Stream 的 MatchResult 对象中）。只有 group( ) 返回的完整匹配是相关的。每个组是一个两行字符串。distinct( ) 删除所有重复的两行字符串，最后 Collector 将 Stream 中的所有字符串连接回一个大 String。

任务 4.1.7：船帆在哪里？

在任何情况下，必须从 Point.Double 对象创建 Stream，最后 Collectors.toMap(...) 创建 Map。有两条路径可以达到目标，因为特殊之处在于，如果将双元素作为键转移到 Map 中，就会出现问题，因为键在 Map 中可能只出现一次。

建议解决方案提出了两种不同的变体。

变体 1：

```
Function<Point.Double, Integer> distanceToCaptain =
    coordinate -> distance( coordinate.x, coordinate.y, 40.2390577, 3.7138939 );

Map<Point.Double, Integer> map =
    Arrays.stream( targets )
          .distinct()
          .collect( Collectors.toMap( Function.identity(), distanceToCaptain ) );
```

列表 4.20 com/tutego/exercise/stream/DistanceToNextStation.java

我们可以使用 distinct( ) 方法轻松删除重复元素，然后 collect(...) 可以使用 Collectors.toMap(...) 将元素传输到 Map。复习一下：

```
Collector<T, ?, Map<K,U>> toMap(Function<? super T, ? extends K> keyMapper,
                               Function<? super T, ? extends U> valueMapper)
```

第一个 Function 确定键，在我们的例子中键是 Stream 中的 Point.Double 对象。Function.identity( ) 是 t –> t，因此 Stream 中的元素也直接是键。第二个函数计算到船长的距离，我们的 Function<Point.Double, Integer> 在内部使用 distance(...)。toMap(...) 就这样建立点和距离之间的关联。

如果没有 distinct( ) 方法，则有两倍的键，并出现异常：IllegalStateException: Duplicate key。

当然也可以没有 distinct( )。

变体 2：

```
map = Arrays.stream( targets )
    .collect( Collectors.toMap( Function.identity(),
    distanceToCaptain, (d,__) -> d ) );
```

列表 4.21 com/tutego/exercise/stream/DistanceToNextStation.java

第二种方法使用了另一个 toMap(...) 方法：

```
Collector<T, ?, Map<K,U>> toMap(Function<? super T, ? extends K> keyMapper,
    Function<? super T, ? extends U> valueMapper,
    BinaryOperator<U> mergeFunction)
```

第三个参数是 BinaryOperator，参数名称已经暴露了它的全部含义：mergeFunction 仅在键出现多次并将值减少为一个结果时才调用。由于在我们的例子中键值对总是相同的，所以我们可以只删除一个坐标距离对。这由 (d,__) -> d 完成；BinaryOperator<Integer> 得到了两个传递的空格；第二个传递被忽略，只返回第一个距离，因为它们总是相同的。

任务 4.1.8：购买最受欢迎的装甲车

Stream API 是 Java 开发人员的福音，但要抵制将所有内容强制转换为 Stream 表达式的诱惑。这项任务就是如此。当其他两个表达式在两分钟内完成它时，在一个难以阅读的 Stream 表达式上投入 30 分钟并不是很经济。因此，提出以下解决方案：

```
Map<String, Long> map1 =
    Arrays.stream( cars )
    .collect( Collectors.groupingBy(
        Function.identity(), Collectors.counting() ) );

map1.entrySet().removeIf( stringLongEntry -> stringLongEntry.getValue() < 2 );
```

列表 4.22 com/tutego/exercise/stream/PopularCar.java

该解决方案使用 groupingBy(...) 方法，该方法提供了作为分类器和关联元素 Collector 的功能。复习一下：

```
Collector<T, ?, Map<K, D>> groupingBy(Function<? super T, ?
    extends K> classifier,
    Collector<? super T, A, D> downstream
```

Map<String, Long> 是通过 Collectors.groupingBy(Function.identity( ), Collectors.counting( )) 形成的。Stream 中的元素形成键，关联的值是该值在 Stream 中出现的次数。

Map 方法 entrySet( ) 返回 Set<Entry<String, Long>>，并且该集合不包含副本，而是数据的实时视图。可以调用 removeIf(Predicate<Entry<String, Long>>)，并且此方法迭代整个集合并准确删除满足传递的谓词的元素。谓词说明应该删除出现次数少于两次的条目。

另一种解决方案使用单独的 Collector，也不再使用 Map<String, Long>，而是 Map<String, Boolean>。

```
Collector<Object, long[], Boolean> collector = Collector.of(
    () -> new long[1],              // Supplier<A> supplier
    (array, string) -> array[0]++, // BiConsumer<A,T> accumulator
    (array1, array2 ) -> { array1[0] += array2[0]; return array1; },
    // BinaryOperator<A> combiner
    array -> array[0] > 1 );        // Function<A,R> finisher

Map<String, Boolean> map2 =
    Arrays.stream( cars ).collect( Collectors.groupingBy(
    Function.identity(), collector ) );
```

列表 4.23 com/tutego/exercise/stream/PopularCar.java

第二种建议解决方案也是使用 groupingBy(Function, Collector)，但是程序没有使用预定义的 Collector，而是编写了自己的 Map<String, Boolean>（用于返回）。可以使用以下静态工厂方法构建 Collector 对象：

```
Collector<T, A, R> of(Supplier<A> supplier,
    BiConsumer<A, T> accumulator,
    BinaryOperator<A> combiner,
```

```
                  Function<A, R> finisher,
                  Characteristics... characteristics)
```

对于所需的返回类型 Map<String, Boolean>，寻找一个返回 Boolean 的 Collector，因此参数化必须如下所示：Collector<Object, XXX, Boolean>。第一种类型可以保留 Object，因为我们不使用字符串。XXX 类型没有出现在 Stream 中，是一个内部容器，我们使用一个 long 数组来存储计数。因此，正确的声明如下：Collector<Object, long[], Boolean>。

在 of(...) 方法中，我们传递了四个必要的参数。在例子中，Characteristics 是可变参数并且无关紧要。

1. Supplier 返回一个长度为一个条目的 long 数组。Collector 记下了其中的数字。Collector 有一个状态，它可以用来稍后确定 String 是否出现了两次以上。
2. BiConsumer 获取 long 数组和 String，但仅递增且不使用 String。因此，类型参数也可以是 Object，而不必是 String。
3. BinaryOperator 连接两个 long 数组。该操作仅在并行流上执行。
4. Collector 处的 Function 将结果映射到 Boolean。每当数组中的计数大于 1 时，结果为 true，从而形成与字符串关联的值。

任务 4.2.1：识别数组中的非数字

```
public static boolean containsNan( double[] numbers ) {
  return DoubleStream.of( numbers ).anyMatch( Double::isNaN );
}
```

列表 4.24 com/tutego/exercise/stream/ArrayContainsNan.java

终端方法 anyMatch(Predicate) 有常规流和原始流，很容易解决这个问题。方法引用 Double::isNaN 是 value –> Double.isNaN(value) 的缩写。

任务 4.2.2：生成数十年

```
public static int[] decades( int start, int end ) {
  return IntStream.rangeClosed( start / 10, end / 10 )
                  .map( x -> x * 10 )
                  .toArray();
}
```

列表 4.25 com/tutego/exercise/stream/DecadesArray.java

原始流 IntStream 和 LongStream 有两个静态 rangeXXX(...) 方法用于创建整数 Stream：

IntStream：
- IntStream range(int startInclusive, int endExclusive);
- IntStream rangeClosed(int startInclusive, int endInclusive)。

LongStream：
- LongStream range(long startInclusive, long endExclusive);
- LongStream rangeClosed(long startInclusive, long endInclusive)。

可以确定开始和结束值，也可以确定结束值是否属于 Stream，但增量始终为 1。由于结束值属于任务中的结果，所以 rangeClosed(...) 是一个不错的选择。

为了解决这个问题，需要将增量从 1 增加到 10。我们能做到这一点，只要：

1. 将 rangeClosed(...) 的开始值和结束值除以 10，得到一个增量 1 的流；
2. 使用 map(...) 将元素乘以 10。

由于目标是一个数组而不是 IntStream，所以 toArray( ) 返回所需的 int[]。

任务 4.2.3：通过 Stream 创建具有恒定内容的数组

```
public static int[] fillNewArray( int size, int value ) {
  if ( size < 0 )
    throw new IllegalArgumentException( "size can not be negative" );

  return IntStream.range( 0, size ).map( __ -> value ).toArray();
}
```

列表 4.26 com/tutego/exercise/stream/GenerateAndFillArray.java

该方法照常检查参数的有效性。大小不能为负数，否则会出现异常。不必检查 value 的分配，因为可以为变量分配任何值。

IntStream.range(...) 生成带有 size 数量元素的 IntStream。在例子中，Stream 生成从 0 到 size 的数字是无关紧要的；我们将所有值转换为固定值。标识符 __ 表示未使用 Lambda 参数。这将创建一个只有常量值的 Stream。toArray(...) 将 Stream 转换为数组。

任务 4.2.4：绘制金字塔（Java 11）

```
IntStream.rangeClosed( 1, 5 )
        .mapToObj( i -> " ".repeat( 5 -i)+ "/\\".repeat(i))
        .forEach( System.out::println );
```

列表 4.27 com/tutego/exercise/stream/StreamPyramid.java

生成的输出具有一个特点，即总是有两个字符∧彼此相邻。第一排是一对，第二排是两对，依此类推。

因此，我们需要创建一个从 1 到所需高度的 IntStream。必须在前面设置的空格也取决于此计数器。对于从 1 到 5 生成的 Stream，空格数是 5 – i。

任务 4.2.5：查找字符串的字母频率

自 Java 8 以来，所有实现 CharSequence 接口的类都有两个新的默认方法：

- IntStream chars( );
- IntStream codePoints( )。

它们返回一个 IntStream，在第一种方法中，Stream 由扩展为 int 值的 char 字符组成；在第二种方法中 Stream 等于 int 值。其区别主要在于复合代码点，对于我们的例子，满足于简单的 chars( ) 方法。

```
String input = "eclectic";
String output =
    input.chars()
    .mapToObj( c -> (char)c + "" +
    input.chars().filter(d -> d == c).count() )
    .collect( Collectors.joining( "/" ) );
System.out.println( output );  // e2/c3/l1/e2/c3/t1/i1/c3
```

列表 4.28 com/tutego/exercise/stream/LetterOccurrences.java

IntStream 为我们提供了一个包含每个字符的 Stream。这些字符现在必须与输入字符串中的相应频率关联。要计算出现次数，可以重复构建一个 IntStream 并使用筛选方法和 count( ) 来查找流中的字符数。如果连接字符和该计数器，则在下一步中我们有一个 Stream<String>，其中每个字符都映射到这个对。最后，这些对必须用斜线连接。使用 Collectors.joining(...) 的精简可以解决这个问题。

任务 4.2.6：从 1 到 0，从 10 到 9

```java
private static String decrementNumbers( Reader input ) {
  return new Scanner( input )
    .findAll( "10|[1-9]" )             // Stream<MatchResult>
    .map( MatchResult::group )         // Stream<String>
    .mapToInt( Integer::parseInt )     // IntStream
    .map( Math::decrementExact )       // IntStream
    .mapToObj( Integer::toString )     // Stream<String>
    .collect( Collectors.joining( ", " ) );
}
```

列表 4.29 com/tutego/exercise/stream/DecrementNumbers.java

通过巧妙地选择 Stream，该任务可以用一个表达式完成。为了从头运行到尾，我们首先看一下各步骤：

1. 将字符串拆分为单独的数字；
2. 减少数字；
3. 将连接的数字转换为字符串。

第一步是识别数字。查找子字符串是正则表达式的任务。正则表达式可以使用 Pattern 类处理，但是这个类对我们处理数据流没有帮助。允许查找与正则表达式匹配的字符串的第二个类是 Scanner。可以将 Scanner 直接应用于 Reader。使用 findAll(...) 方法，Scanner 返回所有发现的 Stream<MatchResult>。正则表达式必须识别所有可能出现的数字，即 10|9|8|…|1。OR 连接可以简写为 10|[1-9]，但不能简写为 [1-9]|10。

将发现作为 MatchResult 返回，我们需要使用 group( ) 访问主组。这给我们返回一个 String。对于加法，使用 mapToInt(...) 将 String 转换为整数。Integer.parseInt(...) 负责处理。可以使用方法引用，因为 Integer.parseInt(...) 匹配 Function 的签名。不会有解析错误，因为正则表达式只匹配数字。

整数本身必须减 1。可以把它写成一个 Lambda 表达式，但是有一个合适的方法也可以使用方法引用来访问：Math.decrementExact(...)。如果超出范围，它会抛出 ArithmeticException，但在我们的例子中不会发生这种情况。

在减少数字之后，所有东西都必须转换为一个大的字符串。这分两步进行。第一步，每个整数被单独转换成一个 String；第二步，这些字符串通过一个特殊的 Collector 被组合成一个长字符串。

任务 4.2.7：合并两个 int 数组

```java
public static final int MAX_ARRAY_LENGTH = Integer.MAX_VALUE - 8;

public static int[] join( int[] numbers1, int[] numbers2, long maxSize ) {
  if ( maxSize > MAX_ARRAY_LENGTH )
     throw new IllegalArgumentException( "Requested array size exceeds VM limit" );

  return IntStream.concat( IntStream.of( numbers1 ), IntStream.of( numbers2 ) )
              .limit( maxSize )
              .toArray();
}

public static int[] join( int[] numbers1, int[] numbers2 ) {
  return join( numbers1, numbers2, (long) numbers1.length + numbers2.length );
}
```

列表 4.30 com/tutego/exercise/stream/JoinIntArrays.java

为了使 join(...) 方法更加灵活，有两种实现方法。join(int[] numbers1, int[] numbers2, int maxSize) 有一个附加参数 maxSize，它限制了数组元素的数量。Java 中没有连接两个数组的方法，但原始流可以解决问题。重点在于方法：

```
IntStream concat(IntStream a, IntStream b)
```

必须用 IntStream.of(...) 从这两个数组中创建一个 IntStream，将它们放入 concat(...)，结果是一个由两个数组组成的原始流。toArray() 最后创建所需的数组。

但是，必须注意一件事：到目前为止，数组不能大于 Integer.MAX_VALUE – 8，这就是为什么建议解决方案中有一个常数。

在 join(...) 方法中，第一个条件判断检查所传递的 maxSize 是否可能比所产生的数组大。即使后面 numbers1.length + numbers2.length > MAX_ARRAY_LENGTH，Stream 对象的 limit（maxSize）产生的数组也保持在界限内。

有两个参数的更简单的 join(...) 方法确定总长度并委托给三个参数方法。数组的长度是 int 类型。我们将其扩展为 long 值，这样总和就是 long 值，也就不会溢出，然后可以看到总和没有超过 MAX_ARRAY_LENGTH。

任务 4.2.8：确定获胜组合

算法如下：声明一个接收字符串的方法。删除该字符串的第一个字符，然后用新创建的字符串递归调用该方法。接着删除第二个字符并再次递归调用该方法，依此类推。当字符串没有更多字母并且为空时，递归方法结束。

刚才描述的方法在建议解决方案 removeLetter(...) 中：

```java
private static void removeLetter( String word, Set<String> hashSet ) {
  if ( word.isEmpty() )
    return;
  words.add( word );
  IntStream.range( 0, word.length() )
      .mapToObj( i -> new StringBuilder( word ).deleteCharAt(
      index ).toString() )
      .forEach( substring -> removeLetter( substring, words ) );
}

public static Set<String> removeLetter( String word ) {
  Set<String> words = new HashSet<>();
  removeLetter( word, words );
  return words;
}
```

列表 4.31 com/tutego/exercise/stream/RemovingLetters.java

产生的字符串进入一个集合，重复的字符串会被移除。由于调用者需要集合并且不应该将空容器传递给方法，所以还有第二个公共方法 removeLetter(String)，它构建了一个 HashSet，用于传递私有方法 removeLetter(String, Set<String >)，最后返还创建的集合。

任务 4.3.1：最快和最慢的桨手

该任务的目标是创建一个 DoubleSummaryStatistics 对象。此统计对象返回有关元素的数量、最小值、最大值和平均值的信息。有两种实现统计对象的方法。

```java
DoubleSummaryStatistics statistics =
    stream.mapToDouble( result -> result.time ).summaryStatistics();

System.out.printf( "count: %d%n", statistics.getCount() );
```

```
System.out.printf( "min:     %.2f%n", statistics.getMin() );
System.out.printf( "max:     %.2f%n", statistics.getMax() );
System.out.printf( "average:%.2f%n", statistics.getAverage() );
```
列表 4.32 com/tutego/exercise/stream/PaddleCompetition.java

第一个变体是使用来自 Stream<Result> 的时间生成一个 DoubleStream，然后在 DoubleStream 上调用 summaryStatistics()。

第二种选择是直接使用相应的 Collector：

```
DoubleSummaryStatistics statistics =
    stream.collect( Collectors.summarizingDouble( result -> result.time ) );
```
列表 4.33 com/tutego/exercise/stream/PaddleCompetition.java

可以给 Collectors.summarizingDouble(...) 一个提取函数，直接从 Stream<Result> 中提取时间；该程序因此节省了一个中间步骤。

任务 4.3.2：计算中位数

```
public static double median( double... values ) {
  if ( values.length < 1 )
    throw new IllegalArgumentException( "array contains no elements" );

  int skip = (values.length - 1) / 2;
  int limit = 2 - values.length% 2;
  return Arrays.stream( values ).sorted().skip( skip )
      .limit(limit ).average().getAsDouble();
}
```
列表 4.34 com/tutego/exercise/stream/DoubleStreamMedian.java

像往常一样，验证输入，如果数组没有元素，则抛出异常。此外，如果 values 为 null，则会出现自动异常。

在计算中位数时需要导航到中间，然后在中间必须考虑一个或两个元素。如果构建 DoubleStream 然后排序，则 skip(...) 允许跳过一定数量的元素，从而到达中间。下一步，limit(...) 将 Stream 中剩余元素的数量减少到 1 个元素（数组有奇数个元素）或 2 个元素（数组有偶数个元素）。

最后，计算平均值，对于 1 个元素没有什么可计算的，但对于 2 个元素，average() 方法返回算术平均值。由于该方法想要一个浮点数作为结果，所以

getAsDouble( ) 返回该数字，这也是有效的，因为 Stream 正好有 1 个元素。另一种 API 设计可以返回 OptionDouble，允许出现 double 数组没有元素的特殊情况。

最激动人心的部分是对移位和极限的计算。为此，该程序引入了两个变量 skip 和 limit，这两个变量是由输入的长度决定的。

举例（见表 4.1）：

表 4.1　limit 和 skip 赋值举例

列表	skip	limit
9,11,11,11,12	2	1
10,10,12,12	1	2

变量初始化如下：

```
int skip  = (values.length - 1) / 2;
int limit = 2 - values.length% 2;
```

这两个变量都取决于数组的长度。中间是数组的长度除以 2，这对于元素数量为奇数的数组来说相当合适。但仅将数组的长度除以 2 会导致元素数量为偶数的数组出现问题。因为在这种情况下，必须考虑中间的左边和中间的右边的两个元素 (这些元素甚至有一个名字，被称为上中值和下中值)。因此，如果在将数组的长度除以 2 之前将长度减少 1，则会在数量为偶数的数组中间之前得到一个元素。

举例（见表 4.2）：

表 4.2　skip 计算举例

列表	(values.length − 1) / 2	(values.length) / 2;
9,11,11,11,12	(5 − 1) / 2 = 2	5 / 2 = 2
10,10,12,12	(4 − 1) / 2 = 1	4 / 2 = 2

对于数组中的偶数个元素，限制必须为 2；对于奇数个元素，限制必须为 1。values.length% 2 对于偶数数量返回 0，对于奇数数量返回 1。因此，表达式 2 − values.length% 2 对于偶数数量返回 2 − 0 = 2，对于奇数数量返回 2 − 1，即 1。

任务 4.3.3：统计温度并绘制图表

该任务分为几种方法。randomTemperaturesForYear(Year) 方法为一年中的每天生成一个随机温度。createRandomTemperatureMap( ) 方法调用 randomTemperatures ForYear(...)5 次。命令行中以小表格形式输出的结果是通过 printTemperatureTable(...) 方法实现的。最后，writeTemperatureHtmlFile(...) 方法将平均值写入文件。

从 randomTemperaturesForYear(Year) 开始：

```java
private static int[] randomTemperaturesForYear( Year year ) {
  int daysInYear = year.length();
  return IntStream.range( 0, daysInYear )
      .mapToDouble( value -> sin( value * PI / daysInYear ) ) // 0..1
      .map( value -> value * 20 ) // 0..20
      .map( value -> value + 10 ) // 10..30
      .mapToInt( value -> (int) (value + 3 * (random() - 0.5)) )
      .toArray();
}

private static SortedMap<Year, int[]> createRandomTemperatureMap() {
  return Stream.iterate( Year.now(), year -> year.minusYears( 1 ) )
            .limit( 5 )
            .collect( toMap( identity(),
            TemperatureYearChart::randomTemperaturesForYear,
            (y1,y2) -> { throw new RuntimeException( "Duplicates" ); },
            TreeMap::new ) );
}
```

列表 4.35 com/tutego/exercise/stream/TemperatureYearChart.java

特殊数据类型 Year 很有用，因为它通过 length( ) 方法返回一年中的天数，因为不是每一年都有 365 天。有了一年中的天数，就可以形成一个 IntStream，然后为一年中的每天生成一个随机温度。平均温度分布显示为一条曲线：年初和年底温度较低，年中温度较高。

这一点可以通过正弦函数很好地表达出来。因此，将整数通过函数转化为正弦值，年初的第一天对应 0 的正弦值，年底的最后一天对应 π(Pi) 的正弦值，即又是 0。中间为正弦峰。0 和 1 之间的正弦值很小，$\sin(\pi)$ 的最大值是 1，因此在下一步，这些值被乘以 20，然后加 10。这些值还不是随机的，因此最后的映射产生随机性，这样正弦值就会有一些上下波动。

createRandomTemperatureMap( ) 使用 randomTemperaturesForYear(...) 方法创建一个有 5 个元素的 Stream<Year>，从当前年份开始，每次向前移动一年。最终的结果是一个 SortedMap<Year, int[]>，其中包括与每年关联的温度值。

然而，不应该使用有两个参数的 toMap(Function keyMapper, Function valueMapper) 方法，因为没有关于 Map 的声明——在内部，OpenJDK 使用 HashMap。在例子中，一个按年份排序的 Map 是很有用的。因此，使用 toMap(Function keyMapper, Function valueMapper, BinaryOperator mergeFunction, Supplier mapFactory)，这样数据就被明确地排序为一个 TreeMap。

这可以进入输出：

```
private static void printTemperatureTable( SortedMap<Year,
    int[]> yearToTemperatures ) {
  yearToTemperatures.forEach( (year, temperatures) -> {
    String temperatureCells =
        Arrays.stream( temperatures )
            .mapToObj( temperature -> String.format( "%2d", temperature ) )
            .collect( Collectors.joining( " | ", "| ", " | " ) );
    System.out.println( "| " + year + " " + temperatureCells );
  } );
}
```

列表 4.36 com/tutego/exercise/stream/TemperatureYearChart.java

printTemperatureTable(...) 负责所有传递年份的表格输出。传递是关联映射，它将年份与温度关联起来。forEach(...) 方法遍历按照年份排序的数据结构，并创建一个字符串 temperatureCells。

为了创建字符串，首先把 int 数组变成 IntStream。每个温度值都被映射到一个字符串，其中在小于 10 的值前面加一个空格，以便输出总是两位数。收集器结合了所有单独的字符串，结果是 temperatureCells。然后，这个温度值的字符串会和前面的年份一起输出到屏幕上。

下面继续讨论统计数据：

```
IntSummaryStatistics yearStatistics =
        Arrays.stream( yearToTemperatures.get( Year.now() ) ).
summaryStatistics();
    System.out.printf( "max:%d, min:%d%n",
```

```
    yearStatistics.getMax(), yearStatistics.getMin() );
```
列表 4.37 com/tutego/exercise/stream/TemperatureYearChart.java

我们已经多次使用 Arrays.stream(...) 方法，它是 IntStream.of(...) 的替代方法，从一个 int 数组创建一个 IntStream。与常规的 Stream 对象相比，这三个原始流有一个特殊的方法 summaryStatistics( )，它提供了一个统计对象，包含了最小值、最大值和平均值的信息。可以很容易地抓住这些信息并输出。

方法 getStatistics(...) 从 Map 中查询 IntSummaryStatistics 的一年中的某个月。

```
private static IntSummaryStatistics getStatistics( YearMonth yearMonth,
    int... temperatures ) {
  int start = yearMonth.atDay( 1 ).getDayOfYear();
  int end   = yearMonth.atEndOfMonth().getDayOfYear();
  return Arrays.stream( temperatures, start - 1, end ).summaryStatistics();
}
```
列表 4.38 com/tutego/exercise/stream/TemperatureYearChart.java

一年中的所有温度值都存储在传递的 int 数组 temperatures 中。如果想计算某个特定月份的统计数据，则必须在适当的位置创建一个子数组。这可以用 Arrays.stream(...) 解决，因为可以给这个数组提供开始和结束索引。关于一年 365 天的问题，例如三月或十二月什么时候，YearMonth 对象提供了答案。对于起始值，用 atDay(1) 得到一个新的 YearMonth 对象作为月初，然后用 getDayOfYear( ) 计算当年的日期。对每个月的最后一天也是如此，使用 atEndOfMonth( ) 方法将 YearMonth 对象设置为月末。我们将起始值和结束值传递给 Arrays.stream(...)，其中月从 1 开始，必须将起始值向左移动一个位置，但不将终点向左移动一个位置，因为值是排他的，而不是包容的。

可以调用如下方法：

```
IntSummaryStatistics monthStatistics =
    getStatistics( YearMonth.of( 2020, SEPTEMBER ),
      yearToTemperatures.get( Year.now() ) );
System.out.printf( "max:%d, min:%d, average:%.2f%n",
    monthStatistics.getMax(),
    monthStatistics.getMin(),
    monthStatistics.getAverage() );
```
列表 4.39 com/tutego/exercise/stream/TemperatureYearChart.java

最后是写入文件的方法：

```
private static void writeTemperatureHtmlFile( Year year, Map<Year, int[]>
yearToTemperatures, Path path ) throws IOException {
  String template =
    "<!DOCTYPE html>\n" +
    "<html lang=\"en\">\n" +
    "<body>\n" +
    "<canvas id=\"chart\" width=\"500\" height=\"200\"></canvas>\n" +
    "<script src=\"https://cdnjs.cloudflare.com/ajax/libs/Chart.js/2.9.3/↩
Chart.bundle.min.js\"></script>\n" +
    "<script>\n" +
    "const cfg = {\n" +
    "    type: ´bar´,\n" +
    "    data: {\n" +
    "        labels: [\"Jan\", \"Feb\", \"Mrz\", \"Apr\", \"Mai\", \"Jun\", " +
    "                 \"Jul\", \"Aug\", \"Sep\", \"Okt\", \"Nov\", \"Dez\"],\n" +
    "        datasets: [ {\n" +
    "            label: \"Average values\", fill: false,\n" +
    "            data: [%s],\n" +
    "        } ]\n" +
    "    },\n" +
    "    options: {\n" +
    "        responsive: true,\n" +
    "        title: { display:true, text:'Temperature curve' },\n" +
    "        tooltips: { mode: 'index', intersect: false },\n" +
    "        hover: { mode: 'nearest', intersect: true },\n" +
    "        scales: {\n" +
    "            xAxes: [ { display: true, scaleLabel: { display: true, labelString: 'Month' } } ],\n" +
    "            yAxes: [ { display: true, scaleLabel: { display: true, labelString: 'Temperature' } } ]\n" +
    "        }\n" +
    "    }\n" +
```

```
"};\n" +
"window.onload = () => ⤸
 new Chart(document.getElementById(\"chart\").getContext(\"2d\"), cfg);\n" +
"</script>\n" +
"</body>\n" +
"</html>";

  String formattedTemperatures =
      IntStream.rangeClosed( JANUARY.getValue(), DECEMBER.getValue() )
        .mapToObj( Month::of )
        .map( month -> year.atMonth( month ) )
        .map( yearMonth ->
          getStatistics( yearMonth, yearToTemperatures.get( year ) ) )
        .map( IntSummaryStatistics::getAverage )
        .map( avgTemperature -> String.format( ENGLISH, "%.1f",
avgTemperature ) )
        .collect( Collectors.joining( "," ) );
  String html = String.format( template, formattedTemperatures );

  Files.write( path, Collections.singleton( html ) );
}
```

列表 4.40 com/tutego/exercise/stream/TemperatureYearChart.java

HTML 文档的声明占据了方法中最大的面积。有一个地方是 data: [%s]，而这个 %s 是格式化字符串中的一个典型占位符，因此稍后可以使用 String.format(...) 方法插入动态计算的值。该方法的工作原理如下：

- rangeClosed(...) 返回一个数字从 1 到 12 的 IntStream。
- mapToObj(...) 将这些数字映射到 Month 对象。
- 年份被传递给 writeTemperatureHtmlFile(...) 方法。表达式 year.atMonth(month) 组合了迭代中的年和月，并为映射返回一个新的 YearMonth 对象。Lambda 表达式 month -> year.atMonth(month) 可以缩写为 year::atMonth，但完整的 Lambda 表达式更具可读性。
- 生成的 Stream<YearMonth> 可以使用自己的 getStatistics(...) 方法映射到 IntSummaryStatistics；结果是 Stream<IntSummaryStatistics> 类型。
- 从这个 IntSummaryStatistics 得到方法引用的平均值，创建了一个 Stream

<Double>。
- 最后格式化这个平均值来创建一个月的字符串。
- 对每个月都这样做，最后收集字符串并用逗号分隔。

字符串 formattedTemperatures 包含现在需要插入 HTML 文档的内容。生成完整的 HTML 文档后，write(...) 写入文件。在 Java 11 之前，只能编写行的集合；从 Java 11 开始也可以编写单个字符串。

writeTemperatureHtmlFile(...) 的应用如下：

```
try {
  Path tempFile = Files.createTempFile( "temperatures", ".html" );
  writeTemperatureHtmlFile( Year.now(), yearToTemperatures, tempFile );
  Desktop.getDesktop().browse( tempFile.toUri() );
}
catch ( IOException e ) { e.printStackTrace(); }
```

列表 4.41 com/tutego/exercise/stream/TemperatureYearChart.java

第 5 章
文件和对文件内容的随机访问

即使大量迁移到云和数据库，文件系统仍然是组织文档的重要存储和场所。Bonny Brain 和 CiaoCiao 船长也在本地存储了很多东西——不应该公开的东西已经够多了。

本章使用的数据类型如下：

- java.io.File (https://docs.oracle.com/en/java/javase/11/docs/api/java.base/java/io/File.html)
- java.nio.file.Path (https://docs.oracle.com/en/java/javase/11/docs/api/java.base/java/nio/file/Path.html)
- java.nio.file.Paths (https://docs.oracle.com/en/java/javase/11/docs/api/java.base/java/nio/file/Paths.html)
- java.nio.file.Files (https://docs.oracle.com/en/java/javase/11/docs/api/java.base/java/nio/file/Files.html)
- java.nio.file.DirectoryStream (https://docs.oracle.com/en/java/javase/11/docs/api/java.base/java/nio/file/DirectoryStream.html)
- java.nio.file.FileVisitor (https://docs.oracle.com/en/java/javase/11/docs/api/java.base/java/nio/file/FileVisitor.html)
- java.io.RandomAccessFile (https://docs.oracle.com/en/java/javase/11/docs/api/java.base/java/io/RandomAccessFile.html)
- java.awt.Desktop (https://docs.oracle.com/en/java/javase/11/docs/api/java.desktop/java/awt/Desktop.html)

5.1 路径和文件

与 Java 中的许多东西一样，处理文件有"旧"和"新"方式。在许多示例中，仍然可以看到 java.io.File，FileInputStream，FileOutputStream，FileReader 和 FileWriter 类型的代码，但这些类型已不再是最新的，这就是在本章中只处理 Path 和 Files 的原因，因为这些类型允许使用虚拟文件系统，如 ZIP 存档。只有当实际涉及本地文件系统的文件或目录时，File 才是强制性的。例如，使用与操作系统关联的程序打开文件或从外部启动的程序重定向数据流。

5.1.1 显示每日格言 ★

CiaoCiao 船长时不时就会失去动力。每天的激励语或格言给脾气暴躁的他带来新的思路。编写一个应用程序，生成一个带有格言的 HTML 文件，然后打开浏览器来显示这个文件。这个任务可以通过 java.nio.files.Files 的两个方法来完成。

任务：
- 使用适当的 Files 方法创建一个以后缀 .html 结尾的临时文件。
- 在新的临时文件中，编写如下 HTML 代码：
  ```
  <!DOCTYPE html><html><body>
  ›Die Dinge, die wir stehlen, sagen uns, wer wir sind.‹
  - Thomas von Tew
  </body></html>
  ```
- 从 java.awt.Desktop 类中找到一个打开默认浏览器并显示 HTML 文件的方法。

5.1.2 合并隐藏 ★

使用某些 Files 方法，可以逐行读取整个文件并再次写入。

CiaoCiao 船长在一个大的文本文件中收集了潜在的隐藏地点，但他经常会不自觉想到更多藏身之处，并迅速把它们写进一个新的文件。现在，他开始花时间清理和总结一切：小的文本文件将与大文件合并。重要的是，不要改变大文件中的条目顺序，只收录小文件中与大文件中不同的条目，因为可能主文件早就包含了这些隐藏。

任务：

编写方法 mergeFiles(Path main, Path... temp)，打开主文件，添加所有临时内容，然后写回主文件。

5.1.3 创建文本副本★★

如果将文件复制到 Windows 资源管理器中的同一文件夹，则会创建一个副本。该副本会自动获得一个新名称。我们需要模拟这种行为的 Java 程序。

任务：

- 编写 Java 方法 cloneFile(Path path) 来创建文件的副本，系统地生成文件名。假设 <Name> 表示文件名，那么第一个副本将称为 Copy of <Name>，随后的文件名将是 Copy(<Zahl>) of <Name>。
- 如果在目录上调用方法或有其他错误，则该方法可能抛出 IOException。

举例：

假设一个文件名为 Top Secret UFO Files.txt。那么新的文件名应该是这样的：

- Copy of Top Secret UFO Files.txt;
- Copy (2) of Top Secret UFO Files.txt;
- Copy (3) of Top Secret UFO Files.txt，等等。

5.1.4 生成目录列表★

用户可以在命令行中显示目录内容和元数据，就像文件选择对话框向用户显示文件一样。

任务：

- 使用 Files 和 newDirectoryStream(...) 方法编写一个程序，列出当前的目录内容。
- 在 DOS 中，启动 dir 程序。完全复制目录列表输出。头和尾不是必需的。

5.1.5 搜索大型 GIF 文件★

Bonny Brain 的硬盘一片混乱，这是因为她将所有图像都保存在同一个目录中。现在找不到上次寻宝的照片了！她只记得这些照片以 GIF 格式保存，宽度超过 1 024 像素。

任务：

给出任意目录。搜索此目录（不是递归！）以查找所有 GIF 格式且最小宽度为 1 024 像素的图像。

使用以下代码读取宽度并检查 GIF 格式：

```java
private static final byte[] GIF87aGIF89a = "GIF87aGIF89a".getBytes();
private static boolean isGifAndWidthGreaterThan1024( Path entry ) {
  if ( ! Files.isRegularFile( entry ) || ! Files.isReadable( entry ) )
    return false;

  try ( RandomAccessFile raf = new RandomAccessFile(entry.toFile(), "r") ) {
    byte[] bytes = new byte[ 8 ];
    raf.read( bytes );
    if ( ! Arrays.equals( bytes, 0, 6, GIF87aGIF89a, 0, 6 ) &&
         ! Arrays.equals( bytes, 0, 6, GIF87aGIF89a, 6, 12 ) )
      return false;

    int width = bytes[ 6 ] + (bytes[ 7 ] << 8);
    return width > 1024;
  }
  catch ( IOException e ) {
    throw new UncheckedIOException( e );
  }
}
```

该方法读取第一批字节并检查前 6 字节是否对应字符串 GIF87a 或 GIF89a。原则上，这个测试也可以用 ! new String(bytes, 0, 6).matches("GIF87a|GIF89a") 实现，但这会涉及内存中的一些临时对象。

检查后，程序读出 2 字节的宽度并将字节转换为 16 位整数。

5.1.6 递归降级目录并找到空的文本文件 ★

Bonny Brain 的硬盘中还有一大堆乱七八糟的东西。由于某种未知原因,她有许多 0 字节的文本文件。

任务:
- 从选定的起始目录开始,通过所有子目录递归运行 FileVisitor,并查找空文本文件。
- 文本文件是具有后缀 .txt(不分大小写)的文件。
- 如果找到文件,则在控制台上显示文件的绝对路径。

5.1.7 开发自己的文件过滤工具库 ★★★

Files 类提供了三种静态方法来查询目录中的所有条目:

- newDirectoryStream(Path dir);
- newDirectoryStream(Path dir, String glob);
- newDirectoryStream(Path dir, DirectoryStream.Filter<? super Path> filter)。

结果始终是 DirectoryStream<Path>。第一种方法不过滤结果,第二种方法允许使用 Glob 字符串,如 *.txt,第三种方法允许任何过滤器。

java.nio.file.DirectoryStream.Filter<T> 是过滤器必须实现的接口。方法是 boolean accept(T entry),它就像一个谓词。

Java 库声明了接口,但没有实现。

任务:
编写 DirectoryStream.Filter 的各种实现,用于检查文件的以下属性:

- 属性(如可读、可写);
- 长度;
- 文件扩展名;
- 使用正则表达式的文件名;
- 魔术初始标识符。

在理想情况下,API 允许连接所有过滤器,如下所示:

```
DirectoryStream.Filter<Path> filter =
```

```
            regularFile.and( readable )
                       .and( largerThan( 100_000 ) )
                       .and( magicNumber( 0x89, 'P', 'N', 'G' ) )
                       .and( globMatches( "*.png" ) )
                       .and( regexContains( "[-]" ) );

try ( DirectoryStream<Path> entries =
  Files.newDirectoryStream( dir, filter ) ) {
    entries.forEach( System.out::println );
}
```

5.2 对文件内容的随机访问

对于文件来说，可以获得一个输入/输出流，并从前到后进行读取或写入。另一个 API 允许随机访问，即位置指针。

5.2.1 输出文本文件的最后一行 ★★

船员们将所有行动记录在电子日志中，并在末尾附加新条目。条目不超过 100 个字符，文本以 UTF-8 编写。

现在 CiaoCiao 船长对最后一个条目感兴趣。如果只从文件中读取最后一行，则 Java 程序会是什么样子？由于日志中已经有很多条目，所以无法读取整个文件。

任务：
- ▶ 编写一个程序，返回文本文件的最后一行。
- ▶ 找到一个不会不必要地占用太多内存的解决方案。

小提示：
思考是否可以合理使用 ([^\r\n]*)$。

5.3 可供参考的解决方案

任务 5.1.1：显示每日格言
```
try {
  String html = "<!DOCTYPE html><html><body>" +
```

```
      ">The things we steal tell us who we are.< - Thomas von Tew" +
      "</body></html>";
    Path tmpPath = Files.createTempFile( "wisdom", ".html" );
    Files.write( tmpPath, Collections.singleton( html ) );
    Desktop.getDesktop().open( tmpPath.toFile() );
  }
  catch ( IOException e ) {
    System.err.println( "Couldn't write HTML file in temp folder or open file" );
    e.printStackTrace();
  }
```

列表 5.1 com/tutego/exercise/io/DailyWordsOfWisdom.java

在建议解决方案中，我们处理三个中心指令。第一步是在临时目录中创建文件。使用 createTempFile(...) 方法可以指定部分名称和后缀，我们选择后缀 .html，以便操作系统稍后可以通过该后缀选择合适的显示程序。createTempFile(...) 返回用来将字符串写入此文件的生成的 Path。

所选择的 write(...) 方法乍一看有点奇怪——因为第二个参数必须是 Iterable<? extends CharSequence>，然而我们只有一个 String。因此，从一个单一的元素生成一个 Collection，从而满足方法的签名。从 Java 11 开始才增加了 writeString(...) 方法，用它可以直接写一个字符串，而不需要绕路。

open(...) 是少数强制需要 File 对象的方法之一。我们从 Path 生成一个 File 对象，用它来打开浏览器，它应该与 HTML 文件的显示相关。另外，可以对网页使用 browse(URI)，并通过 toUri() 从 Path 中获取 URI。

任务 5.1.2：合并隐藏

```
public static void mergeFiles( Path main, Path... temp ) throws IOException {

  Iterable<Path> paths =
      Stream.concat( Stream.of( main ), Stream.of( temp ) )::iterator;
  Collection<String> words = new LinkedHashSet<>();

  for ( Path path : paths )
    try ( Stream<String> lines = Files.lines( path ) ) {
      lines.forEach( words::add );
    }
```

```
    Files.write( main, words );
}
```

列表 5.2 com/tutego/exercise/io/MergeFiles.java

对于此任务，数据结构 LinkedHashSet 是最适合的，因为作为一个集合，它只包含一次元素，并且考虑了插入元素的顺序。只需要确保第一个文件的行首先被插入数据结构，然后是其他文件的行。

为了读入这些行并将其插入数据结构，第一个文件的处理方式与其他文件相同。但是，这种统一只可能是绕道实现，因为参数列表中的第一个数据类型是一个单一的 Path 变量，然后跟随一个可变参数，也就是一个 Path 数组。建议解决方案首先将第一个元素放在一个 Stream 中，然后将其与可变参数数组中的第二个元素 Stream 结合起来，结果是一个 Stream<Path>。我们只需要运行它。原则上，forEach(...) 在这里是个不错的选择，但有一个问题：输入/输出操作会触发已检查的异常，而这与 Lambda 表达式不兼容。因此，Stream 被转换为 Iterable，以便可以利用扩展的 for 循环。方法引用 Iterator 返回一个 Iterable 类型的表达式。这是一个不错的技巧，因为 Stream 本身并没有实现 Iterable。

扩展的 for 循环遍历文件，读取所有行，将它们放入数据结构，最后将所有行写回第一个文件。

任务 5.1.3：创建文本副本

```
private static final String COPY_OF          = "Copy of %s";
private static final String NUMBERED_COPY_OF = "Copy (%d) of %s";

public static void cloneFile( Path path ) throws IOException {

  if ( Files.isDirectory( path ) )
    throw new IllegalArgumentException(
      "Path has to be a file but was a directory" );

  Path parent   = path.getParent();
  Path filename = path.getFileName();

  Path copyPath = parent.resolve( String.format( COPY_OF, filename ) );

  for ( int i = 2; Files.exists( copyPath ); i++ )
```

```
      copyPath = parent.resolve( String.format( NUMBERED_COPY_OF, i,
filename ) );
```

```
    Files.copy( path, copyPath );
  }
```
列表 5.3 com/tutego/exercise/io/FileClone.java

解决方案中的算法可以这样的方式进行：依次生成可能的文件名，并进行测试，直到找到一个空文件名。圆括号里有一个计数器（从 2 开始），没有 Copy (1) of <Name>。

方法 cloneFile(Path path) 开始时查询是否有一个目录被错误地传递为 Path，在这种情况下抛出一个异常。我们不能克隆目录。如果是一个文件，就提取该文件的目录和文件名。

可能的新文件名的第一个样本以 Copy of 开头，还不包含计数器。可以用 Files.exists(...) 测试这个文件名是否存在。如果该文件存在，则必须继续使用计数器。因此，我们将这个存在性测试设置为 for 循环中的一个条件，并使用一个计数器变量 i，在开始时将其初始化为 2，以便能够用圆括号表示该计数器。在该循环的主体中，变量 copyPath 被重新赋值，总是用圆括号内的循环计数器进行赋值。运行循环，直到找到一个不存在的 copyPath，然后循环结束，Files.copy(...) 根据由 copyPath 指定的路径创建一个文件的副本。

为了方便改变不同语言的字符串，它们被提取为常数。

任务 5.1.4：生成目录列表

```
private final static DateTimeFormatter ddMMyyyy_hhmm =
    DateTimeFormatter.ofPattern( "dd.MM.yyyy hh:mm" );

static void listDirectory( Path dir ) throws IOException {
  try ( DirectoryStream<Path> entries = Files.newDirectoryStream( dir ) ) {
    for ( Path path : entries ) {
      Instant instant = Files.getLastModifiedTime( path ).toInstant();
      LocalDateTime dateTime = LocalDateTime.ofInstant(
        instant, ZoneId.systemDefault() );
      String formattedDateTime = dateTime.format( ddMMyyyy_hhmm );
      String dirLength = Files.isDirectory( path ) ? "<DIR> :
        String.format( "%,14d", Files.size( path ) );
      String filename = path.getFileName().toString();
```

```
      System.out.printf( "%s %s %s%n",
        formattedDateTime, dirLength, filename );
    }
  }
}
```
列表 5.4 com/tutego/exercise/io/DosDir.java

不同的 API 一起完成了这个任务。我们需要类 Files、类型 Path、日期 / 时间计算和格式字符串。

由于文件操作可以抛出潜在的异常，但无法处理这些异常，所以我们的方法将把潜在的异常传递给调用者。我们还使用了 try-with-resources——资源是 DirectoryStream。如果代码既快又脏，则会经常把扩展的 for 循环冒号右侧的 DirectoryStream 看成 Iterable。DirectoryStream 是需要关闭的资源。用于遍历目录中所有条目的扩展的 for 循环可以在下一步找到。

变量 Path 现在包含可以代表文件或目录的路径。无论如何都要查询上次访问的时间。Files.getLastModifiedTime(...) 确实返回了必要的 FileTime 对象，但 toString( ) 方法没有返回相应结果。因此，为了获得良好的输出，需要"绕个小弯路"：首先，将 FileTime 带入 Instant，然后将其转换为 LocalDateTime，使用格式为 "dd.MM.yyyy hh:mm" 的 DateTimeFormatter，已经为该格式引入了自己的常量。

有了日期和时间，现在其他部分也随之而来。根据路径是目录还是文件，我们必须设置 <DIR>，或者如果是文件，则询问文件长度；String.format(...) 使字节数达到适当的长度。

最后，询问文件或目录的名称，并将所有内容放在同一行。这一行以格式化的日期和时间开头，然后是对是否为目录的说明（否则是文件长度），最后是文件名或目录名。

任务 5.1.5：搜索大型 GIF 文件
该解决方案使用 Files 方法：

```
DirectoryStream<Path> newDirectoryStream(
  Path dir, DirectoryStream.Filter<? super Path> filter)
```

对我们而言，DirectoryStream.Filter<? super Path> filter 是相关的，因为它可用于实现限制结果的标准。让我们看一下相关 UML 图示（见图 5.1）。

Filter 是 DirectoryStream 的嵌套类型。如果需要传递一个 Filter，则需要实现 Filter 接口和 accept(...) 方法。可以通过类、Lambda 表达式或方法引用来完成实现，并且 boolean isGifAndWidthGreaterThan1024(Path entry) "凑巧" 匹配 boolean accept(Path entry)，这正是方法引用。

```
Path directory = Paths.get( name );
try {
  try ( DirectoryStream<Path> files = Files.newDirectoryStream( directory,
      FindBigGifImages::isGifAndWidthGreaterThan1024 ) ) {
    files.forEach( System.out::println );
  }
}

catch ( IOException e ) {
  e.printStackTrace();
}
```
列表 5.5 com/tutego/exercise/io/FindBigGifImages.java

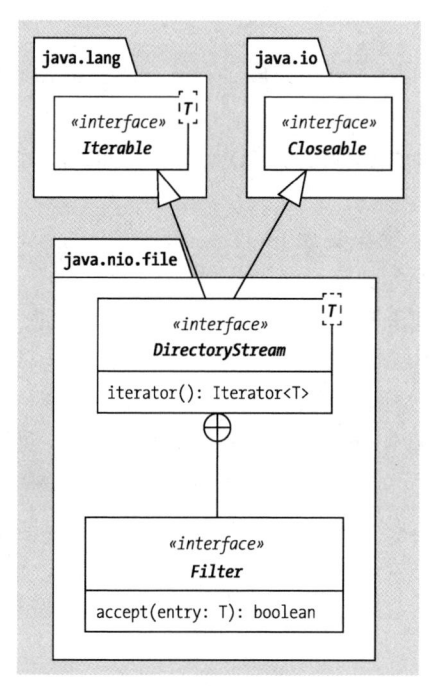

图 5.1 DirectoryStream 和附属类型的 UML 图示

Files.newDirectoryStream(...) 方法返回 DirectoryStream，它是 AutoCloseable 并且必须在最后再次关闭。这可以通过 try-with-resources 方便地处理。DirectoryStream

也是一个 Iterable，因此可以在扩展的 for 循环中或使用 forEach(...) 和对应的 Consumer 来运行它。

任务 5.1.6：递归降级目录并找到空的文本文件

```java
public static void findEmptyTextFiles( Path base, Consumer<Path> callback )
    throws IOException {
  class PrintingFileVisitor extends SimpleFileVisitor<Path> {
    @Override
    public FileVisitResult visitFile( Path visitedFile,
                                      BasicFileAttributes fileAttributes ) {
      if ( visitedFile.toString().toLowerCase().endsWith( ".txt" )
          && fileAttributes.size() == 0L )
        callback.accept( visitedFile );
      return FileVisitResult.CONTINUE;
    }
  }
  Files.walkFileTree( base, new PrintingFileVisitor() );
}
```

列表 5.6 com/tutego/exercise/io/EmptyFilesFinder.java

该解决方案使用两个参数实现 findEmptyTextFiles(...) 方法：第一个参数用于基本目录，第二个参数用于找到路径时调用的使用者。

静态方法 Files.walkFileTree(...) 在文件系统中递归地从一个目录运行到所需的深度。在例子中，不限制深度，也不提供任何其他选项。该方法必须传递一个 FileVisitor 的实现。这不是函数式接口，而是四方法接口。可以自己实现该接口，但 Java 库使用 SimpleFileVisitor 提供了一个简单的实现（见图 5.2）。

形成这个类的一个子类并覆盖与我们相关的 visitFile(...) 方法，只要 walkFileTree(...) 找到一个文件就会调用该方法。walkFileTree(...) 不允许过滤器，这很遗憾，因为 newDirectoryStream(...) 允许过滤器。因此，必须实现文件名以 .txt 结尾且大小为 0 字节的标准。谢天谢地，通过 visitFile(...) 得到了：

1. 传递路径，以便可以检查文件名是否以 .txt 结尾；
2. 属性，可以用它们来测试文件是否为空。

如果两个条件都符合，则调用回调函数并将路径传递给文件，然后继续在目录中搜索。为此，方法返回 FileVisitResult.CONTINUE。

```
┌─ java.nio.file ─────────────────────────────────────────────────────┐
│  «enumeration»      ┌──────────────────«interface»──────────────────┐│
│  FileVisitResult    │               FileVisitor                    ││
│  ─────────────      │ preVisitDirectory(dir: T, attrs: BasicFileAttributes): FileVisitResult ││
│  CONTINUE           │ visitFile(file: T, attrs: BasicFileAttributes): FileVisitResult ││
│  TERMINATE          │ visitFileFailed(file: T, exc: IOException): FileVisitResult ││
│  SKIP_SUBTREE       │ postVisitDirectory(dir: T, exc: IOException): FileVisitResult ││
│  SKIP_SIBLINGS      └──────────────────────△───────────────────────┘│
│                                            ┆                         │
│                     ┌──────────────────────┴───────────────────────┐│
│                     │              SimpleFileVisitor               ││
│                     │ preVisitDirectory(dir: T, attrs: BasicFileAttributes): FileVisitResult ││
│                     │ visitFile(file: T, attrs: BasicFileAttributes): FileVisitResult ││
│                     │ visitFileFailed(file: T, exc: IOException): FileVisitResult ││
│                     │ postVisitDirectory(dir: T, exc: IOException): FileVisitResult ││
│                     └──────────────────────────────────────────────┘│
└────────────────────────────────────────────────────────────────────┘
```

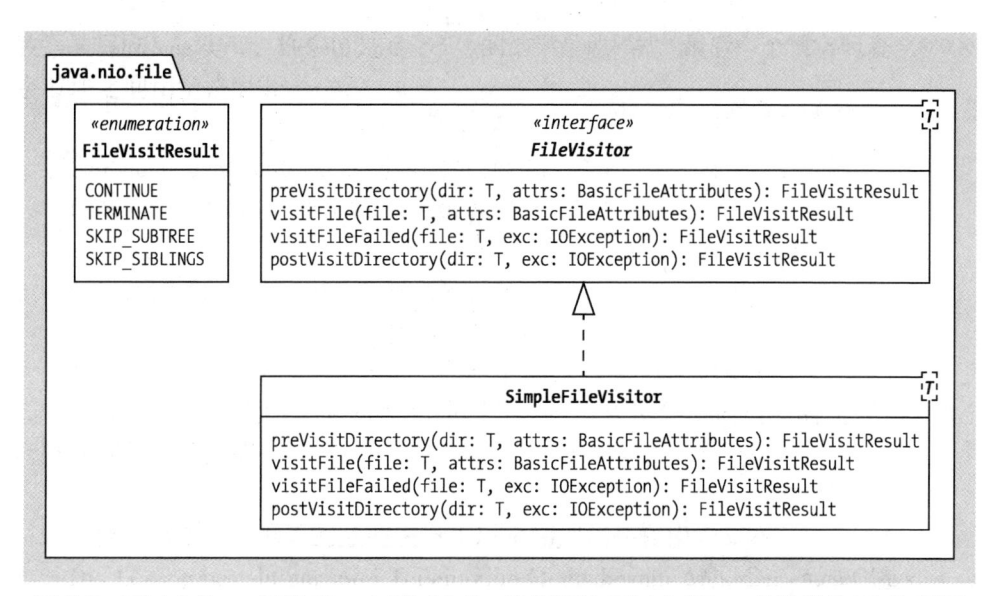

图 5.2　FileVisitor、子类 SimpleFileVisitor 和返回的 FileVisitResult 枚举的 UML 图示

任务 5.1.7：开发自己的文件过滤工具库

```java
class FileFilters {

  public interface AbstractFilter extends DirectoryStream.Filter<Path> {
    default AbstractFilter and( AbstractFilter other ) {
      return path -> accept( path ) && other.accept( path );
    }

    default AbstractFilter negate() {
      return path -> ! accept( path );
    }

    static AbstractFilter not( AbstractFilter target ) {
      return target.negate();
    }
  }

  /**
   * Tests if a {@code Path} is readable.
   */
```

```java
    public static final AbstractFilter readable = Files::isReadable;

    /**
     * Tests if a {@code Path} is writable.
     */
    public static final AbstractFilter writable = Files::isWritable;

    /**
     * Tests if a {@code Path} is a regular file.
     */
    public static final AbstractFilter directory = Files::isDirectory;

    /**
     * Tests if a {@code Path} is a regular file.
     */
    public static final AbstractFilter regularFile = Files::isRegularFile;

    /**
     * Tests if a {@code Path} is hidden.
     */
    public static final AbstractFilter hidden = Files::isHidden;

    /**
     * Tests if the file size of a {@code Path} is zero.
     */
    public static final AbstractFilter empty = path -> Files.size( path ) == 0L;

    /**
     * Tests if the file size of a {@code Path} is larger than the specified size.
     */
    public static AbstractFilter largerThan( long size ) {
        return path -> Files.size( path ) > size;
    }
```

```java
/**
 * Tests if the file size of a {@code Path} is smaller than the specified
 * size.
 */
public static AbstractFilter smallerThan( long size ) {
    return path -> Files.size( path ) < size;
}

/**
 * Tests if a {@code Path} is older than the specified {@code FileTime}.
 */
public static AbstractFilter olderThan( FileTime other ) {
    return path -> Files.getLastModifiedTime( path ).compareTo( other ) > 0;
}

/**
 * Tests if a {@code Path} has a specified suffix, ignoring case, e.g. ".txt".
 */
public static AbstractFilter suffix( String suffix, String... more ) {
    return path ->
        Stream.concat( Stream.of( suffix ), Stream.of( more ) )
            .anyMatch( aSuffix -> {
                String filename   = path.toString();
                int suffixLen     = aSuffix.length();
                int suffixOffset  = filename.length() - suffixLen;
                return filename.regionMatches( /* ignore case */ true,
                    suffixOffset, suffix, 0, suffixLen );
            } );
}

/**
 * Tests if the content of a {@code Path} starts with a specified sequence of
 * bytes.
 */
public static AbstractFilter magicNumber( int... bytes ) {
    ByteBuffer byteBuffer = ByteBuffer.allocate( bytes.length );
```

```java
      for ( int b : bytes ) byteBuffer.put( (byte) b );
      return magicNumber( byteBuffer.array() );
    }

    /**
     * Tests if the content of a {@code Path} starts with a specified sequence of
     * bytes.
     */
    public static AbstractFilter magicNumber( byte... bytes ) {
      return path -> {
        try ( InputStream in = Files.newInputStream( path ) ) {
          byte[] buffer = new byte[ bytes.length ];
          in.read( buffer );
          // If file is smaller than bytes.length, the result is false
          return Arrays.equals( bytes, buffer );
        }
      };
    }

    /**
     * Tests if a {@code Path} regexContains a specified regex.
     */
    public static AbstractFilter regexContains( String regex ) {
      return path -> Pattern.compile( regex ).matcher( path.toString() ).find();
    }

    /**
     * Tests if a filename of a {@code Path} matches a given glob string.
     */
    public static AbstractFilter globMatches( String glob ) {
      return path -> path.getFileSystem().getPathMatcher( "glob:" + glob )
                         .matches( path.getFileName() );
    }
```

列表 5.7 com/tutego/exercise/io/FileFiltersDemo.java

Files.newDirectoryStream(…) 方法允许递归地运行目录树。该方法需要函数式接口 Filter 及其 accept(...) 方法的实现作为传输：

```
boolean accept(T entry) throws IOException
```

该接口没有任何其他 default 或 static 方法。此外，类型参数对我们来说始终是 Path，因此想创建一个新接口 AbstractFilter 作为 Filter 的子类型，其中包含两个额外的 default 方法和一个 static 方法。and(...) 方法用逻辑 and 连接两个 AbstractFilters，negate( ) 方法否定它自己的 accept(...) 方法的结果，静态 not(...) 方法返回一个新的 AbstractFilter 对象，它也否定了结果。

FileFilters 类声明了各种常数和方法。每当有东西需要被参数化时，就会使用一种方法；如果没有必要进行参数化，则用一个常数就足够了。以下所有的常量都是 AbstractFilter 数据类型，并且这些方法都返回 AbstractFilter。常量通过方法引用 Files 类中的相应方法进行初始化。

让我们专注于更有趣的方法。

- hasSuffix(...) 将所有可能的文件后缀连接到一个 Stream 中，然后查询路径是否有传递的文件后缀之一。为此，首先将路径转换为字符串，并通过 regionMatches (...) 测试后缀，不区分大小写。使用 regionMatches(...) 的测试比 toLowerCase(...) 和 endsWith(...) 更让人困惑。但 regionMatches(...) 不会创建临时对象，而且性能更好。

- magicNumber(byte...) 需要一个可变的字节数，因此参数变量为 null 会导致 NullPointerException，否则：

1. 会打开一个输入流；
2. 读入的字节数正好是参数数组的大小；
3. 两个数组用 Arrays.equals(...) 进行比较。如果文件小于传递的字节数，那么 equals(...) 方法在任何情况下都会返回 false，因为该方法首先检查相同的元素数。

- magicNumber(int...) 是 magicNumber(byte...) 的重载变体，参数列表中的字节是不灵活的，因为其值范围是 −128～+127。例如开发人员必须编写：
magicNumber((byte) 0x89, (byte) 'P', (byte) 'N', (byte) 'G')
int 类型对调用者来说更方便：
magicNumber(0x89, 'P', 'N', 'G')
因此，magicNumber(int...) 将 int[] 转换为 byte[] 并委托给 main(...) 方法。

- regexContains(...) 接收一个正则表达式字符串，它被编译，然后应用于路径，如果 find() 方法返回 true，则文件名与正则表达式匹配。
- globMatches(...) 检查 Glob 字符串的文件名；这些是简单的表达式，如 *.txt 或 ??-??-1976。getPathMatcher(...) 返回带有前缀 glob: 或 regex: 的 PathMatcher。regexContains(...) 的实现不使用带有正则表达式的 PathMatcher，因为 PathMatcher 测试是否完全匹配，但解决方案需要部分搜索。

任务 5.2.1：输出文本文件的最后一行

```java
private static final int MAX_LINE_LENGTH = 100;
private static final int MAX_NUMBER_OF_BYTES_PER_UTF_8_CHAR = 4;

private static void printLastLine( String filename ) throws IOException {
  try ( RandomAccessFile file = new RandomAccessFile( filename, "r" ) ) {
    int blockSize = MAX_LINE_LENGTH * MAX_NUMBER_OF_BYTES_PER_UTF_8_CHAR;
    file.seek( file.length() - blockSize );
    byte[] bytes = new byte[ blockSize ];
    file.read( bytes );

    String string = new String( bytes, StandardCharsets.UTF_8 );
    Matcher matcher = Pattern.compile( "([^\\r\\n]*)$" ).matcher( string );
    if ( matcher.find() )
      System.out.println( matcher.group( 1 ) );
  }
}
```
列表 5.8 com/tutego/exercise/io/LastLine.java

　　解决方案包括两个步骤：首先，在文件末尾读入一个块；然后，从该块中提取最后一行。

　　为了确定正确的块大小，将任务已知的最大行长度 (100) 乘以每个 UTF-8 字符预期的最大字节数。在 UTF-8 编码中，最多 4 个字符可以编码 1 个符号。当最大行长度未知时，任务变得更加困难，但是乘积 MAX_LINE_LENGTH * MAX_NUMBER_OF_BYTES_PER_UTF_8_CHAR 告诉我们，文件中这个大小的最后一个块也包含最后一行。

　　下一步，将文件指针设置为最后一个块的开头。读入一个字节数组，并根据 UTF-8 编码将其转换为字符串。如果最后一行没有最大长度，则读取的块包含前一行的剩余部分或带有行尾字符的行和最后一行。为了提取最后一行，可以使用

lastIndexOf(...)，但是正则表达式 ([^\\r\\n]*)$ 返回的最后一行更好。正则表达式的组成如下：

1. 末尾的美元符号表示输入结束；
2. 括号中的组代表要提取的内容；
3. 在字符类 [^\\r\\n] 中，小帽子否定所有非行尾字符，带星号的 [^\\r\\n]* 提供了一系列此类非行尾字符。

如果有匹配，则输出它，否则没有输出。

第 6 章
输入 / 输出

本章主要介绍数据流，也就是数据的连续流动。它们可以通过多个过滤器写入目标或从源头读取。Java 的输入 / 输出嵌套是一个体现抽象和灵活性的非常好的例子，有助于我们自己的过滤器建模。

本章使用的数据类型如下：

- java.io.OutputStream (https://docs.oracle.com/en/java/javase/11/docs/api/ java.base/java/io/OutputStream.html)
- java.io.InputStream (https://docs.oracle.com/en/java/javase/11/docs/api/ java.base/java/io/InputStream.html)
- java.io.Reader (https://docs.oracle.com/en/java/javase/11/docs/api/java.base/ java/io/Reader.html)
- java.io.Writer (https://docs.oracle.com/en/java/javase/11/docs/api/java.base/ java/io/Writer.html)
- java.nio.files.Files (https://docs.oracle.com/en/java/javase/11/docs/api/ java.base/java/nio/file/Files.html)
- java.util.Scanner (https://docs.oracle.com/en/java/javase/11/docs/api/java.base/ java/util/Scanner.html)
- java.lang.Readable (https://docs.oracle.com/en/java/javase/11/docs/api/ java.base/java/lang/Readable.html)
- java.lang.Appendable (https://docs.oracle.com/en/java/javase/11/docs/api/ java.base/java/lang/Appendable.html)
- java.io.IOException (https://docs.oracle.com/en/java/javase/11/docs/api/ java.base/java/io/IOException.html)

- java.io.DataInputStream (https://docs.oracle.com/en/java/javase/11/docs/api/java.base/java/io/DataInputStream.html)
- java.io.DataOutputStream (https://docs.oracle.com/en/java/javase/11/docs/api/java.base/java/io/DataOutputStream.html)
- java.io.FilterInputStream (https://docs.oracle.com/en/java/javase/11/docs/api/java.base/java/io/FilterInputStream.html)
- java.io.FilterOutputStream (https://docs.oracle.com/en/java/javase/11/docs/api/java.base/java/io/FilterOutputStream.html)
- java.io.FilterReader (https://docs.oracle.com/en/java/javase/11/docs/api/ java.base/java/io/FilterReader.html)
- java.io.FilterWriter (https://docs.oracle.com/en/java/javase/11/docs/api/ java.base/java/io/FilterWriter.html)
- java.io.DataInput (https://docs.oracle.com/en/java/javase/11/docs/api/java.base/ java/io/DataInput.html)
- java.io.DataOutput (https://docs.oracle.com/en/java/javase/11/docs/api/ java.base/java/io/DataOutput.html)
- java.util.zip.GZIPOutputStream (https://docs.oracle.com/en/java/javase/11/docs/api/java.base/java/util/zip/GZIPOutputStream.html)
- java.io.ByteArrayOutputStream (https://docs.oracle.com/en/java/javase/11/docs/ api/java.base/java/io/ByteArrayOutputStream.html)
- java.io.InputStreamReader (https://docs.oracle.com/en/java/javase/11/docs/api/java.base/java/io/InputStreamReader.html)
- java.io.OutputStreamWriter (https://docs.oracle.com/en/java/javase/11/docs/api/ java.base/java/io/OutputStreamWriter.html)
- java.io.PrintWriter (https://docs.oracle.com/en/java/javase/11/docs/api/ java.base/java/io/PrintWriter.html)
- java.lang.AutoCloseable (https://docs.oracle.com/en/java/javase/11/docs/api/java.base/java/lang/AutoCloseable.html)
- java.io.BufferedOutputStream (https://docs.oracle.com/en/java/javase/11/docs/api/java.base/java/io/BufferedOutputStream.html)
- java.util.zip.CheckedOutputStream (https://docs.oracle.com/en/java/javase/11/docs/api/java.base/java/util/zip/CheckedOutputStream.html)
- java.io.Serializable (https://docs.oracle.com/en/java/javase/11/docs/api/ java.base/java/io/Serializable.html)

6.1 直接数据流

在 Java 中有 4 种不同直接数据流：InputStream 和 OutputStrea(面向字节的读和写)，以及 Reader 和 Writer（面向字符的读和写）。首先开始的任务是直接向资源中写入流或直接从资源中读取流。

6.1.1 确定不同位置的数量（读取文件）

CiaoCiao 船长得到两个文本文件，它们乍看之下相同，但是他想确认这个两个文件是否完全匹配或有差异。

任务：
- 编写一个返回不同位置的数量的方法 long distance(Path file1, Path file2)，在计算机术语中它被称作汉明距离（英文：Hamming distance）。
- 可以假设这两个文件的长度完全相同。

举例：
一个文件包含的字符串如下：

```
To Err is Human. To Arr is Pirate.
```

另外一个文件包含的字符串如下：

```
To Arr is Human. To Err is Pirate!
```

距离为 3，因为 3 个符号不匹配。

6.1.2 将 Python 程序转换为 Java 程序（文件写入）★

在第 2 章中，解决了将 SVG 输出写入屏幕的各种问题——现在直接将其输出写入 HTML 文件。提醒一下，以下 HTML 包含一个 SVG，其中包含一个高度和宽度为 1 和 x-y 坐标的矩形（10×10）：

```
<!DOCTYPE html>
<html><body>
```

```
<svg width="256" height="256">
 <rect x="10" y="10" width="1" height="1" style="fill:rgb(0,29,0);" />
</svg>
</body></html>
```

在一本关于计算机生成艺术的书中，CiaoCiao 船长在前几页找到了一幅插图。该样品是由 Python 程序编写的：

```
import Image, ImageDraw
image = Image.new("RGB", (256, 256))
drawingTool = ImageDraw.Draw(image)

for x in range(256):
    for y in range(256):
        drawingTool.point((x, y), (0, x^y, 0))

del drawingTool
image.save("xorpic.png", "PNG")
```

Python 函数 point(...) 获得 x-y 坐标和 RGB 颜色信息。其中，三个参数 0，x^y，0 分别代表红、绿、蓝三个分量。

任务:
- 因为 CiaoCiao 船长不喜欢蛇，所以必须将 Python 程序转换成 Java 程序。
- 最终结果应该是一个带有 SVG 块的 HTML 文件，而不是 PNG 文件，其中，每个像素是一个 1×1 的 SVG 矩形。

奖励:
最后用浏览器打开 HTML 文件，Desktop-Klasse 将在这里提供帮助。

6.1.3 生成目标代码（写入文件）★

CiaoCiao 船长从邮局收到越来越多带有粉红色条形码的信件。他首先想到了这是来自 Bonny Brain 的编码爱情信息，但随后意识到信封上有一个所谓的目标代码——这是邮政编码。

竖破折号中的数字和编码见表 6.1，其中下划线 "_" 象征间距用一个空格表示。

表 6.1　数字和编码

数字	编码
0	| | | |
1	| | | _
2	| | _ |
3	| | _ _
4	| _ | |
5	| _ | _
6	| _ _ |
7	_ | _ |
8	_ | | |
9	_ | | _

任务：

- 编写一个静态方法 writeZielcode(String, Writer)，将含有指定编码的带数字的字符串写入一个 Writer。
- 一个数字的 4 个符号之间应该有 2 个空格。

举例：
字符串"023"会在文件中被写成"ⅠⅠⅠⅠ ⅠⅠ Ⅱ"。

> **提示：**
> 从 Files 中获取一个 Writer, 以便写入文件。
> 更多的关于编码的信息可以参见网页 https://de.wikipedia.org/wiki/Zielcode。

6.1.4 将文件内容转换为小写字母（文件的读取和写入）

从一种格式到另一种格式的文本转换是常见的操作。

任务：
打开一个文本文件，读取每个字符，并将其转换为小写，然后将其写入一个新文件。编写一个执行此操作的方法并将其称为 convertFileToLowercase(Path inPath, Path outPath)。

6.1.5 以 ASCII 灰度显示 PPM 图像 ★★★

由于格式不同，所以生成的像素图形总是有点复杂。但是，有一种非常简单的基于 ASCII 的文件格式，称为 PPM (Portable Pixel Map)。其规范很简单（http://netpbm.sourceforge.net/doc/ppm.html）。

一个 Java 程序可以很容易地生成 PPM 图像。然而，它在 Windows 操作系统中的一个缺点是需要第三方程序进行显示，例如免费软件 GIMP (https://www.gimp.org)。

以下例子显示了 PPM 文件的基本结构：

```
P3
3 2
255
255   0   0
  0 255   0
  0   0 255
255 255   0
255 255 255
  0   0   0
```

有各种空格分隔的标记。我们制定了以下规则:

- 第一个标记是标识符 P3;
- 接下来是图像的宽度和高度;
- 下面是最大的颜色数字,其总是被假设为 255;
- 所有像素的红、绿、蓝值从左上到右下依次排列;
- 高度和宽度以及颜色的数值总是正数。

任务:
- 读入 PPM 文件并提取所有颜色值。
- 将每个颜色值传递到灰度值。
- 图像的每个点都要成为一个 ASCII 字符。将 0~255 的每个灰度值转换成 ASCII 字符。
- 允许在程序中把 RGB 值转换为灰度值并参数化,这样就可以交换算法。
- 允许从灰度值到 ASCII 字符的参数化转换。

以下接口和常数可用于灰度值转换:

```
public interface RgbToGray {
  RgbToGray DEFAULT = (r, g, b) -> (r + g + b) / 3;
  int toGray( int r, int g, int b );
}
```

通过使用 IntBinaryOperator,Java 提供了从 (int, int) 到 int 的映射,但没有具有 3 个参数的功能接口。

平均方法是高效的,但不符合人类的感知。更逼真的插图考虑到普通人对颜色的感知不同。众所周知的是高度方法,其简单的平均值方法是 0.21 R + 0.72 G + 0.07 B。

IntUnaryOperator 接口适合用于将一个灰度值(int)映射到一个 ASCII 字符(char,扩展为 int)。默认转换器如下所示:

```
public enum GrayToAscii implements IntUnaryOperator {
  DEFAULT;

  private final char[] ASCII_FOR_SHADE_OF_GRAY =
    // black = 0, white = 255
```

```
    "@MBENRWDFQASUbehGmLOYkqgnsozCuJcry1v7lit{}?j|()=~!-/<>\"^_';,:`. "
      .toCharArray();
  private final int CHARS_PER_RGB = 256 / ASCII_FOR_SHADE_OF_GRAY.length;
  @Override public int applyAsInt( int gray ) {
    return ASCII_FOR_SHADE_OF_GRAY[ gray / CHARS_PER_RGB ];
  }
}
```

默认字符串 1 的长度[①]为 64 个字符。原则上，这意味着黑色成为 @，白色成为一个空格。

举例：
上面 PPM 文件的结果如下：

```
kkk
? @
```

6.1.6 文件分块处理（文件的读和写）

在 Anaa-Atoll，港口软件已经在 Commodore PC-30 上运行了大约 40 年。Bonny Brain 已经成功地操纵了这台计算机，但现在软件需要更新，必须通过软盘导入。3.5 英寸高清软盘在默认情况下可以存储 1 474 560 字节（1 440 KB）内容。软件更新不适合在软盘中进行，因此需要有软件将一个大文件（"兼容磁盘"）分割成几个小文件。

任务：
编写一个程序，在命令行中获取文件名，然后将该文件拆分为几个较小的部分。

列举：
调用如下所示：

```
$ java com.tutego.exercise.io.FileSplitter Hanjaab.bin
```

如果文件 Hanjaab.bin 的大小是 2.440 KB，那么 Java 程序会把它分成 Hanjaab.bin.1（1 440 KB）和 Hanjaab.bin.2（1 000 KB）。

[①] 该字符是 https://www.pouet.net/topic.php?which=8056&page=1. 的简化。

第 6 章　输入/输出 | 221

6.2 嵌套流

流可以像俄罗斯套娃一样嵌套。一个流是其核心的真正资源，而其他流像壳一样包裹在它的周围。通过对外壳的操作最终会进入核心。

6.2.1 测试 DataInputStream 和 DataOutputStream

DataInputStream 和 DataOutputStream 是装饰器，分别增强了简单的 InputStream 和 OutputStream。

- java.io.DataInputStream 和 java.io.DataOutputStream 是如何实施的？可以通过开发环境或者网页 https://github.com/openjdk/jdk/blob/master/src/ java.base/share/classes/java/io/DataInputStream.java 和 https://github.com/open- jdk/jdk/blob/master/src/java.base/share/classes/java/io/DataOutputStream.java 了解。

- java.io.FilterInputStream (https://docs.oracle.com/en/ java/javase/11/docs/api/java.base/java/io/FilterInputStream.html) 和 java.io. FilterOutputStream (https://docs.oracle.com/en/java/javase/11/docs/api/java.base/ java/io/FilterOutputStream.html) 的任务是什么？

6.2.2 使用 GZIPOutputStream 压缩数字序列 ★

java.util.zip.GZIPOutputStream (https://docs.oracle.com/en/java/javase/11/docs/ api/java.base/java/util/zip/GZIPOutputStream.html) 是一个特殊的输出流，可以无损失地压缩数据。

任务：

- 创建一个数字从 0 到 N 的压缩文件，使用 writeLong(…) 将其写入 GZIPOutput Stream。
- 比较不同 N 的文件尺寸。
- 从哪个 N 开始是值得压缩的？

6.3 序列化

通过序列化，Java 使得将对象的状态写入数据流，随后从数据流中重新分配对象成为可能，这个过程被称为反序列化。

ObjectOutputStream 和 ObjectInputStream 类用于将 Java 对象转换成二进制流，反之亦然。所有要被序列化的引用类型必须是可序列化的。我们想在接下来的任务中使用这些类型，并了解序列化的实际例子。

这两个类都是典型的装饰器：当序列化时，ObjectOutputStream 确定数据并将串行字节序列写入构造函数中传递的 OutputStream；当读入时，情况正好相反，这里 ObjectInputStream 从传递的 InputStream 中读取。

6.3.1 对聊天的数据进行（反）序列化并转换为文本★★

用一个聊天程序来传输 Java 对象，但是聊天程序只能传输 ASCII 字符。因此，不仅必须对对象进行（反）序列化，还必须将其转换为文本格式和从文本格式转换回对象。

任务：

- 编写一个方法 objectToBase64(Object) 来序列化一个对象，然后用 DeflaterOutputStream 压缩，最终返回一个 Base64 编码。
- 编写一个方法 deserializeObjectFromBase64(String)，将 Base64 编码的字符串放入字节流，将其与 InflaterInputStream 解包，并将其用作反序列化的源。

> **提示：**
> Base64.Encoder 和 Base64.Decoder，特别是 die wrap(…) 方法可以帮助将一个字符串转换成二进制数据（或反过来）。

6.3.2 测试：序列化的前提条件★

如果从下面的 Inputs 类形成一个对象，那么可以对 ObjectOutputStream 序列化吗？或者说可能不符合哪些先决条件？

```
class Inputs {
  public static class Input {
    String input;
  }
  public List<Input> inputs = new ArrayList<>();
}
```

6.3.3 保存最后的条目 ★★

Bonny Brain 经常使用 STRING2UPPERCASE 的应用程序，它的核心看起来是这样的：

```
for ( String line; (line = new Scanner( System.in ).nextLine()) != null; )
    System.out.println( line.toUpperCase() );
```

但是，现在每个用户的输入都要被保存在文件系统中，以便在启动应用程序时显示输入。

任务:
- 为项目的所有输入设置以下容器：
```
class Inputs implements Serializable {
  public static class Input implements Serializable {
    String input;
  }
  public List<Input> inputs = new ArrayList<>();
}
```
- 无论何时，一旦有用户输入，这个输入就会被加入一个 Inputs 对象。
- 每次输入后，Inputs 要被序列化为一个文件。
- 当应用程序重新启动时，所有序列化的值应在开始时显示在屏幕上。由于不存在的文件或错误的文件格式引起的异常可以被记录，所以其应被忽略。
- 在 Input 中，将对象变量 input 的数据类型 String 更改为数据类型 CharSequence。重新启动程序。当 Inputs 被反序列化时会发生什么？有问题吗？
- 在 Inputs 和 Input 中设置行：
```
private static final long serialVersionUID = 1;
```
- 重新启动程序，并且序列化新的数据。
- 在 Input 加入一行：
```
LocalDateTime localDateTime = LocalDateTime.now();
```
对于附加对象变量，重启程序，发生了什么或没有发生什么？

6.4 建议解决方案

任务 6.1.1：确定不同位置的数量（读取文件）
```
public static long distance( Path file1, Path file2 ) throws IOException {

  long filesize1 = Files.size( file1 );
  long filesize2 = Files.size( file2 );
```

```java
    if ( filesize1 != filesize2 )
      throw new IllegalStateException(
        String.format( "File size is not equal, but%d for%s and%d for%s",
          filesize1, file1, filesize2, file2 ) );
    long result = 0;

    try ( Reader input1 = Files.newBufferedReader( file1 );
          Reader input2 = Files.newBufferedReader( file2 ) ) {

      for ( int i = 0; i < filesize1; i++ )
        if ( input1.read() != input2.read() )
          result++;
    }

    return result;
  }
```
源代码 6.1 com/tutego/exercise/io/HammingDistance.java

一个重要的先决条件是文件大小相等。因此，该程序首先查询文件大小并比较，如果它们不匹配，就会出现一个 IllegalStateException。这个报错信息非常精确并会显示文件及其大小，这样一个外行人也很容易追查到这个错误。

如果一切顺利，则在下一步为这两个文件建立两个资源。我们为 Reader 使用 Files 方法 newBufferedReader(…)。使用这个方法有两个原因：首先我们想处理字符串而不是二进制流，因此是 Reader 而不是 InputStream；其次，出于性能原因，缓冲很重要，newBufferedReader(...) 返回具有内部缓冲区的 Reader。从内部缓冲区读取单个字符，并且每个字符都无法访问文件系统，这会很慢。

由于已经知道字符的数量，所以会运行一个循环并从每个流中获取一个符号。如果符号不匹配，就递增一个计数器，并在最后返回。

Try-mit-Ressourcen 再次关闭两个流，即使在处理过程中也应该有一个错误。该方法不处理异常，而是将其传递给调用者。在请求文件大小时、打开文件时和读取字符时都可能出现错误。

任务 6.1.2：将 Python 程序转换为 Java 程序（文件写入）

```java
final String filename = "xorpic.html";
try {
```

```
try ( Writer out = Files.newBufferedWriter( Paths.get( filename ) );
      PrintWriter printer = new PrintWriter( out ) ) {

  printer.println( "<!DOCTYPE html>" );
  printer.println( "<html><body><svg width=\"256\" height=\"256\">" );

  for ( int x = 0; x < 256; x++ )
    for ( int y = 0; y < 256; y++ )
      printer.printf(
          "<rect x=\"%d\" y=\"%d\" width=\"1\" "+
          "height=\"1\" style=\"fill:rgb(0,%d,0);\" />",
          x, y, x ^ y );

  printer.println( "</svg></body></html>" );
  }
  Desktop.getDesktop().open( new File( filename ) );
}
catch ( IOException e ) {
  e.printStackTrace();
}
}
```

源代码 6.2 com/tutego/exercise/io/XorFractal.java

Java 和 Python 是差异很大的编程语言，它们的库也不相同。因此，它们的代码没有什么共同点，几乎所有的东西都是不同的。

在 Java 中有很多种写入文件的可能性，通常的类有 FileOutputStream, FileWriter, PrintWriter, Formatter。省略了基于 OutputStream 的类型，因为我们不想写字节，而是写 Unicode 字符。由于格式化字符串相当有用，所以 FileWriter 被省略了，Formatter 也被省略了，因为它只能写格式化字符串，而不是简单的没有格式化的字符串。

由于在输入/输出操作中总是会出现问题，Java 方法会抛出异常，所以我们必须处理这些异常。这由第一个 try 块负责，它会捕获每一个 IOException。

这个特殊的区块 (Block) 没有 catch 分支，因此 try-mit-Ressourcen 块只应该在最后关闭资源，错误处理由外部 try-catch 块负责。

首先建立一个 BufferedWriter，然后用 PrintWriter 来装饰它，这样就有一个用于写格式化的字符串的方法。

下一步是编写 HTML 文件的序言。在两个嵌套循环中，printf(...) 将 SVG 矩形写入数据流。Python 中的 3 个值是 RGB 的颜色值，红色和蓝色分量为 0，即未赋值。该程序仅将绿色分量写入 x 和 y 坐标的 XOR。x 和 y 的取值范围是 0～255，这也恰好是 8 位 RGB 颜色值的最大值。

在运行完两个循环后，try-mit-Ressourcen 块关闭了已打开的流。这两个 try 块乍看之下是奇怪的嵌套，这是因为在成功写入结束后，要用浏览器打开文件。然而，我们必须考虑两个特殊性。try-mit-Ressourcen 块必须首先写入并关闭文件，然后才允许我们再次打开文件进行查看。只有在文件确定没有错误写入的情况下才允许打开文件。如果写入时出错，则不能打开文件。这就是这两个嵌套写作块的逻辑。

任务 6.1.3：生成目标代码（写入文件）

```java
private static final String[] ZIELCODE = {
    "||||",      // 0000 = 0
    "||| ",      // 0001 = 1
    "|| |",      // 0010 = 2
    "||  ",      // 0011 = 3
    "| ||",      // 0100 = 4
    "| | ",      // 0101 = 5
    "|  |",      // 0110 = 6
    " | |",      // not 0111 = 7 but 1010 = 10
    " |||",      // 1000 = 8
    " || " };    // 1001 = 9

public static void writeZielcode( String string, Writer writer )
  throws IOException {
  for ( int i = 0; i < string.length(); i++ ) {
    int value = Character.getNumericValue( string.charAt( i ) );
    if ( value >= 0 && value <= 9 ) {
      writer.write( ZIELCODE[ value ] );
      if ( i != string.length() - 1 )
        writer.write( " " );
    }
  }
}
```

源代码 6.3 com/tutego/exercise/io/Zielcode.java

为了完成这个任务，需要逐个字符地运行一个字符串，并将字符映射到符号序列上。

我们可以有不同的方法。例如，可以将数字与 switch-case 进行比较，然后将相应的字符串写入 Writer。另一个解决方案是联想存储器，可以事先用目标代码的复合字符建立。这里显示的拟议解决方案使用一个数组，其中的条目与该位置的相应目标代码完全对应，switch-case 可以直接对 char 进行大小写区分，但是对于数组上的索引，需要数值；Character.getNumericValue(...) 可以帮助解决这个问题。这种方法的最大优点是所有数字适用于所有语言。一个有效的结果是在 0~9 的数值范围内，有了这个数值，就可以访问数组，然后将数值写入 Writer。如果还没有到达输入字符串中的最后一位，则写入两个空格作为分隔符。

这个清单包含了对数组的注释，很好地说明了竖折号和空格符原则上无非是二进制数字的表示法。一个特殊数字是 7，它不是用我们预计的位模式 0111 表示的，而是用 1010 表示的，即映射到符号 _ | _ |；然而，1010 将是数字 10 的位模式。如果 | _ _ 代表 7，就会涉及太多空白，下划线又象征空格符，这会使读者困惑。

如果把数字解释为一个位模式，那么就可以编写一个稍微不同的不需要数组的解决方案：

```
String string = "0123456789";
for ( int i = 0; i < string.length(); i++ ) {
  BigInteger v = new BigInteger(
    string.charAt( i ) == '7' ? "10" : string.substring( i, i + 1 ) );
  System.out.print( v.testBit( 3 ) ? ' ' : '|' );
  System.out.print( v.testBit( 2 ) ? ' ' : '|' );
  System.out.print( v.testBit( 1 ) ? ' ' : '|' );
  System.out.print( v.testBit( 0 ) ? ' ' : '|' );
  System.out.print( " " );
}
```

源代码 6.4 com/tutego/exercise/io/Zielcode.java

像往常一样，我们运行字符串，首先检查在这个位置是否有一个 7。如果有，则将该数字转移到字符串"10"；如果该数字不是 7，则用 substring(…) 切出一个长度为 1 的字符串（与字符一样）。这两种情况下的结果都是一个字符串。这个字符串用于在 BigInteger 的构造函数中进行初始化。BigInteger 有一个实用的方法 testBit(...)，它以 true 或 false 回答一个位是否被设置。只需要查询 3，2，1 和 0 位，

(检查这个位是一个空格还是一个竖条)。若结果和任务要求不一样,则直接将结果输出在屏幕上。

任务 6.1.4:将文件内容转换为小写字母(文件的读取和写入)

```java
private static final int EOF = -1;

static void convertFileToLowercase( String source, String target )
  throws IOException {
  convertFileToLowercase( Paths.get( source ), Paths.get( target ) );
}

static void convertFileToLowercase( Path source, Path target )
  throws IOException {

  try ( BufferedReader reader = Files.newBufferedReader( source );
        BufferedWriter writer = Files.newBufferedWriter( target ) ) {
    for ( int c; (c = reader.read()) != EOF; )
      writer.write( Character.toLowerCase( (char) c ) );
  }
}
```

源代码 6.5 com/tutego/exercise/io/ConvertFileToLowercase.java

建议解决方案首先声明一个私有静态变量 EOF,稍后将使用这个变量,因为我们正在逐个字符地运行文件,并且 –1 表示流中没有字符。

实际的方法 convertFileToLowercase(...) 被重载,一次是参数类型为 String,另一次是参数类型为 Path。带有文件名的变体创建 Path 对象,并将实际转换委托给第二个方法。

可以通过 Files-Methoden 请求 Reader 和 Writer,给出输入文件和输出文件的路径。这两个对象都有一个很好的特性,即它们都是自动进行缓冲的,因此逐个字符的处理比 Reader 和 Writer 不进行缓冲时快很多。读取时,BufferedReader 首先创建一个 8 KB 大小的缓冲区,然后将其填充到最大值。首先从这个缓冲区中读取单个字符。这同样适用于写入:在内部缓冲区内收集所有数据,当缓冲区满的时候,BufferedWriter 将缓冲区的数据写入底层输出流。

for 循环声明一个变量 c 是要读的字符。在 for 循环的条件表达式中,程序首先读取一个字符,并将结果分配给变量 c,然后将其与 EOF 进行比较。只要读取字符,循环就会运行。在循环体中,将字符转换为大写并写入 Writer。

任务 6.1.5：以 ASCII 灰度显示 PPM 图像

```java
class PPM {

  public interface RgbToGray {
    RgbToGray DEFAULT = (r, g, b) -> (r + g + b) / 3;
    int toGray( int r, int g, int b );
  }

  public enum GrayToAscii implements IntUnaryOperator {
    DEFAULT;

    private static final char[] ASCII_FOR_SHADE_OF_GRAY =
      // black = 0, white = 255
      "@MBENRWDFQASUbehGmLOYkqgnsozCuJcry1v7lit{}?j|()=~!-/<>\"^_';,:'. "
      .toCharArray();
    private static final int CHARS_PER_RGB =
      256 / ASCII_FOR_SHADE_OF_GRAY.length;
    @Override public int applyAsInt( int gray ) {
      return ASCII_FOR_SHADE_OF_GRAY[ gray / CHARS_PER_RGB ];
    }
  }

  private static final String MAGIC_NUMBER = "P3";

  private PPM() { }

  private static String nextStringOrThrow( Scanner scanner, String msg ) {
    if ( ! scanner.hasNext() )
      throw new IllegalStateException( msg );

    return scanner.next();
  }

  private static int nextIntOrThrow( Scanner scanner, String msg ) {
    if ( ! scanner.hasNextInt() )
      throw new IllegalStateException( msg );
```

```java
    int number = scanner.nextInt();
    if ( number < 0 )
      throw new IllegalStateException(
        "Value has to be positive but was " + number );
    return number;
}

public static void renderP3PpmImage( Readable input, RgbToGray rgbToGray,
    IntUnaryOperator grayToAscii, Appendable output )
      throws IOException {

    Scanner scanner = new Scanner( input );

    // Header P3
    String magicNumber = nextStringOrThrow( scanner,
      "End of file, missing header" );
    if ( ! magicNumber.equals( MAGIC_NUMBER ) )
      throw new IllegalStateException( "No P3 image file, but " +
        magicNumber );

    // Width Height
    int width = nextIntOrThrow( scanner,
      "End of file or wrong format for width" );
    int height = nextIntOrThrow( scanner,
      "End of file or wrong format for height" );

    // Max color value
    int maxVal = nextIntOrThrow( scanner,
      "End of file or wrong format for max value" );
    if ( maxVal != 255 )
      throw new IllegalStateException(
        "Only the maximum color value 255 is allowed but was " + maxVal );

    // Matrix
    for ( int y = 0; y < height; y++ ) {
      for ( int x = 0; x < width; x++ ) {
```

```java
      int r = nextIntOrThrow( scanner,
        "End of file or wrong format for red value" );
      int g = nextIntOrThrow( scanner,
        "End of file or wrong format for green value" );
      int b = nextIntOrThrow( scanner,
        "End of file or wrong format for blue value" );
      int gray = rgbToGray.toGray( r, g, b );
      output.append( (char) grayToAscii.applyAsInt( gray ) );
    }
    output.append( '\n' );
  }
}

  public static void renderP3PpmImage( Readable input, Appendable output )
    throws IOException {
    renderP3PpmImage( input, RgbToGray.DEFAULT, GrayToAscii.DEFAULT, output );
  }
}
```

源代码 6.6 com/tutego/exercise/io/PPM.java

由于该类只有静态方法，所以不需要构造函数，并且它被设置为私有。该类不存储自己的对象状态。

Scanner 类非常适合删除连续的标记。可能出现两种类型的错误。数据流中可能缺少数据，或者数据类型不正确。两个辅助方法 nextStringOrThrow(...) 和 nextIntOr- Throw(...) 简化了字符串或整数的读取，并在数据流中没有标记时抛出一个异常。用于读取整数的方法还会检查该数字是否错误地为负值，如果是，则也会抛出一个 IllegalStateException。

从外部访问有两个重载方法 renderP3PpmImage(...)。首先从完整的方法开始，它期望有 4 个转移：

1. 一个来自 Readable 的输入；
2. 一个 RgbToGray 类型的映射，用于将 RGB 值映射到灰度值；
3. 一个 IntUnaryOperator 类型的映射，用于将灰度值转换为 ASCII 值；
4. 一个 appendable 作为输出目标，用于写入结果。

该方法可以读取一个可能的 IOException，像往常一样，在读写过程中可能出现输入/输出错误。

Scanner 和 Readable 相连,也和来源相连,它可以从中读取数据。它获取一个标记并期待一个特殊的标头 P3。这是方法 nextStringOrThrow(...) 的唯一用途。

读取标头后,高度和宽度也必须被读取。它们不能是负数;但是,赋值是 0 也不是错误,要允许这样。然后,读入可能的最大颜色值,根据定义,它必须总是 255。原则上,标准允许任意的值,但我们将其简化了。

一旦有了高度和宽度,就可以编写两个嵌套循环,每个嵌套循环读取三种色调。

原则上,一个循环就足够了,但程序可能希望以后使用点的 x-y 坐标。读取 RGB 值后,调用转换器函数并创建灰度色调,然后映射为 ASCII 字符。IntUnaryOperator 的结果是一个 int 值。

我们将其转换为 char 值并写入输出流。在行尾写一个换行符。

第二个重载方法 renderP3PpmImage(...) 使用两个映射的标准实施。库的用户可以选择是否使用标准转换器或传递他们自己的映射。

任务 6.1.6:文件分块处理(文件的读和写)

```java
private static final int EOF = -1;

private static void splitFile( Path source, int size ) throws IOException {
  Objects.requireNonNull( source );
  Objects.checkIndex( size, Integer.MAX_VALUE );

  try ( InputStream fis = Files.newInputStream( source ) ) {
    byte[] buffer = new byte[ size ];
    for ( int cnt = 1, remaining; (remaining = fis.read( buffer )) != EOF;
      cnt++ )
      try ( OutputStream fos = Files.newOutputStream( Paths.get( source +
          "." + cnt ) ) ) {
        fos.write( buffer, 0, remaining );
      }
  }
}

public static void main( String[] args ) {

  if ( args.length == 0 ) {
    System.err.println( "You need to specify a file name to split the file." );
```

```
      return;
    }

    try {
      String filename = args[ 0 ];
      splitFile( Paths.get( filename ), 1_474_560 );
    }
    catch ( IOException e ) {
      System.err.println( e.getMessage() );
    }
  }
}
```
源代码 6.7 com/tutego/exercise/io/FileSplitter.java

当我们稍后使用 read(...) 方法时,如果无法读取新字节,则它将返回 –1。为此,引入一个常数 EOF。

可以使用 splitFile(...) 方法获取文件的路径和大小。路径可能为零,索引可能为负。因此,稍后会抛出异常,但我们想事先检查正确性。这里我们转向 Objects 类的两个静态方法。

如果下面有输入/输出异常,但 splitFile(...) 没有捕获到这些异常,而是将它们向上传递,那么应该如何处理? 在打开文件、读取内容和写入时都可能出现异常。

第一步,打开文件进行读取。由于要逐个处理文件,所以调用 InputStream。这个资源在程序结束时必须再次关闭,因此使用 try-mit-Ressourcen 块。这个分解可以通过两种不同的方式实现:一种方式是打开 OutputStream,从 InputStream 中读取字节并将其写入 OutputStream,这将非常节省内存。这里选择了另一种方式,以节省程序代码,但却承担了 OutOfMemoryError 的风险,因为该方式一次就读取了整个字节数组,而这个数组正好和传递的大小一样大。然而,不要期待软盘的预期容量,读入缓冲区并直接写入会带来良好的性能。

数组的大小是字节大小。在循环中反复使用这个数组。在循环中,我们声明了两个变量,一个用于生成文件后缀的计数器,另一个用于实际从输入流中读取的字节数。实际的读取是在 for 循环的条件下实现的,在读取之后得到更新的剩余变量,它要么是 –1,要么包含实际读取的字节数。

在循环体中,字节数组被写入文件。Files.newOutputStream(…) 返回一个输出流。该方法的第一个参数是一个生成的 Path-Objekt, 它接收文件名,在它后面添加一个计数器。在 OutputStream 上,write(byte[], int, int) 将部分数组写入文件。如果循环多次运行,则字节数组保证在倒数第二次运行时被完全填满。在最后一次运行中,预测没有 size 个字节被读取,因此更少数组的字节被写入,最后总是

remaining <= size。

main(...) 方法查询是否在命令行中传递了参数，如果是，则调用 fileSplit(...). 将异常写入错误输出通道。

测试 6.2.1: DataInputStream 和 DataOutputStream

FilterInputStream, FilterOutputStream FilterReader 以及 FilterWriter 等类都是拥有自己的过滤器的有用的超类。一方面，它们将封装好的对象存储在一个对象变量中，另一方面，它们有助于确保不是每个来自 InputStream，OutputStream，Reader 或 Writer 的方法都必须被重新执行。为了更详细地了解这一点，看一下 FilterOutputStream 的执行。

```java
package java.io;

public class FilterOutputStream extends OutputStream {
    protected OutputStream out;
    // some fields omitted

    public FilterOutputStream(OutputStream out) {
        this.out = out;
    }

    @Override
    public void write(int b) throws IOException {
        out.write(b);
    }

    @Override
    public void write(byte b[]) throws IOException {
        write(b, 0, b.length);
    }

    @Override
    public void write(byte b[], int off, int len) throws IOException {
        if ((off | len | (b.length - (len + off)) | (off + len)) < 0)
            throw new IndexOutOfBoundsException();
```

```
        for (int i = 0 ; i < len ; i++) {
            write(b[off + i]);
        }
    }

    // flush() / close() omitted
}
```

源代码 6.8 来自 FilterOutputStream 中 OpenJDK 执行的部分

以上程序代码清楚地表明，只有 write(int) 方法转移到封装好的流，其他两个 write(...) 方法只调用 write(int)。自己的过滤器只需要覆盖 write(int)。然而，出于性能方面的考虑，执行其他方法也是有意义的，因为写入整个字节数组要比对数组中的每个元素进行单一的写访问快。

DataInputStream 和 DataOutputStream 是特别的过滤类。在 UML 图（见图 6.1）中 DataInput-Stream 的方法层有更详细的显示；此外，它们还运行了 DataInput 和 DataOutput。

抽象超类 InputStream 和 OutputStream 只适用于字节和字节数组，就像超类 Reader 和 Writer 只适用于数据类型 char、char-Array、String 或 CharBuffer。DataInputStream 和 DataOutputStream 类的特殊之处在于它们也为其他原始数据类型提供了方法。整数或浮点数都可以读/写。这是一个典型的装饰器的例子，它提供了一个更强大的 API，并在后台使用简单的方法。readInt() 的执行是一个很好的例子，它可说明这一点。

```java
public final int readInt() throws IOException {
    int ch1 = in.read();
    int ch2 = in.read();
    int ch3 = in.read();
    int ch4 = in.read();

    if ((ch1 | ch2 | ch3 | ch4) < 0)
        throw new EOFException();
    return ((ch1 << 24) + (ch2 << 16) + (ch3 << 8) + (ch4 << 0));
}
```

源代码 6.9 从 DataInputStream 来的 readInt() 的 OpenJDK- 执行

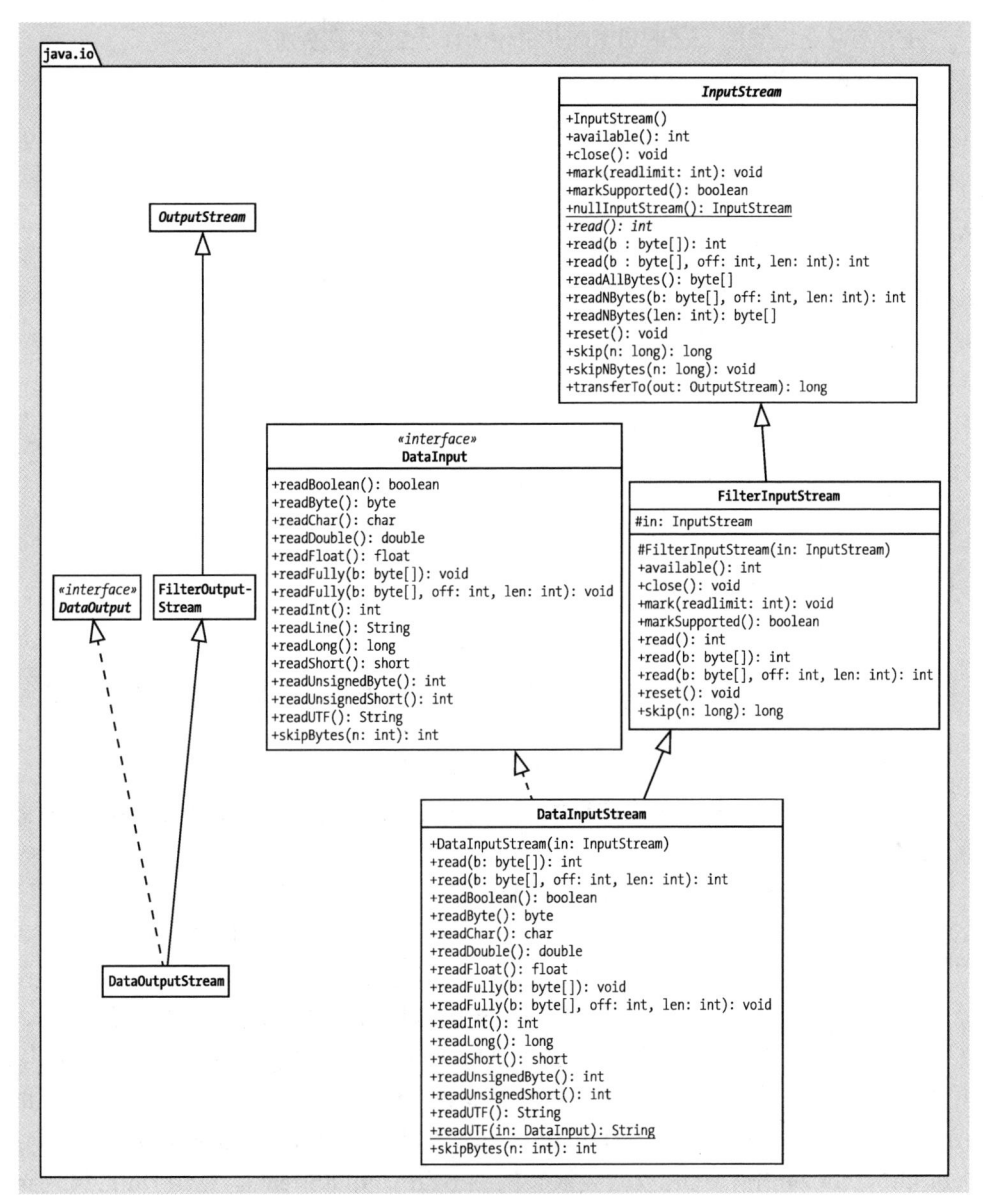

图 6.1 ULM 图

DataInputStream 是一个 FilterInputStream，它在受保护的变量中引用 DataInput Stream 封装的数据流。要读取一个整数，必须从底层资源中读取 4 个单独的字节。在下一步中，必须对 int 值的各字节进行定位和添加。非常巧妙的方法是查询其中一个 read(...) 操作是否导致 –1，这将意味着数据流的结束。单个的返回之间是以"或"的方式联系在一起的，如果其中一个值是负的，则整体结果也会变成负的。

任务 6.2.2：使用 GZIPOutputStream 压缩数字序列

```
Path tempFile = Files.createTempFile( "numbers", "bin.Z" );

final int n = 4;

try ( OutputStream      fos = Files.newOutputStream( tempFile );
      OutputStream      gos = new GZIPOutputStream( fos );
      DataOutputStream  out = new DataOutputStream( gos ) ) {
  for ( int i = 0; i < n; i++ )
    out.writeLong( i );
}

System.out.println( "Uncompressed: " + n * Long.BYTES );
System.out.println( "Compressed:" + Files.size( tempFile ) );

Files.delete( tempFile );
```

源代码 6.10 com/tutego/exercise/io/CompressLotOfNumbers.java

在这个例子中，不在当前目录中而在操作系统的临时目录中创建文件。

在临时目录中创建的文件应该被定期删除，在 Java 中也尝试在程序结束时删除临时文件。

在初始化常数 n 之后，可以很容易地在以后的例子中修改，我们建立了 3 个相互嵌套的流。使用 try-with-resources 块，因为这些流都是 AutoCloseable 类型。嵌套流就像嵌套的环，最里面的一环包含了输出流，它可以写入文件。围绕它的是一个压缩流：写入 GZIPOutputStream 的任何内容都被压缩，然后写入文件流。最后一个环是一个带有强大 API 的装饰器。因此，try-with-resources 块的数据类型不再是 OutputStream，而是 DataOutputStream，因为它有想要的 writeLong(long)-Methode。DataOutputStream 将自身包裹在压缩数据流周围。在 DataOutputStream 中写入的数据会被转发到 GZIPOutputStream。GZIPOutputStream 引导数据进入要写入的输出流。

在循环中，写入 n 个数字。最后，我们计算当数据没有被压缩时文件会有多大。

为此，可以轻松地计算出文件的大小，而不需要创建一个实际文件。为了得到文件压缩后的大小，我们使用 Files.size(...) 方法；另一种解决方案是计算同时流过流的字节数——但我们在这里没有这样做。

未经压缩的文件大小将是 8 000 000 字节，压缩后的大小是 2 129 303 字节。

数据被写成 long 类型，许多位是 0。产生的结果是位模式（最后一个数字代表 999 999，字节块之间用空格隔开）：

```
00000000 00000000 00000000 00000000
00000000 00000000 00000000 00000001
...
00000000 00001111 01000010 00111110
00000000 00001111 01000010 00111111
```

压缩后的文件大约是原始大小的四分之一。从 4 个 long 类型的元素起，这个特殊序列的压缩是值得的。

任务 6.3.1：对聊天的数据进行（反）序列化并转换为文本

Die Lösung besteht aus zwei Teilen: eine Methode zum Abbilden eines Objektes auf einen String und eine Abbildung einer String-Repräsentation zurück auf ein Objekt. Beginnen wir bei der Abbildung auf einen String:

该解决方案由两部分组成：一个将对象映射到字符串的方法和一个字符串表示回到一个对象的映射。让我们从对字符串的映射开始：

```java
public static String serializeObjectToBase64( Object object ) {
  ByteArrayOutputStream baos = new ByteArrayOutputStream();

  try ( OutputStream b64os    = Base64.getEncoder().wrap( baos );
        OutputStream dos= new DeflaterOutputStream( b64os );
        ObjectOutputStream oos = new ObjectOutputStream( dos ) ) {
    oos.writeObject( object );
  }
  catch ( IOException e ) {
    throw new IllegalStateException( e );
  }

  try {
    return baos.toString( StandardCharsets.US_ASCII.name() );
  }
  catch ( UnsupportedEncodingException e ) {
```

```
      throw new IllegalStateException( e );
    }
  }
```
源代码 6.11 com/tutego/exercise/io/ObjectBase64.java

ObjectOutputStream 将字节写入 DeflaterOutputStream，后者会压缩字节，然后将它们传递给 Base64 类 OutputStream。该 API 有异常，因为在通常情况下，目标是在类的构造函数中指定的；但对于 Base64 编码器和解码器来说，它会使用方法 wrap(...)。Base 64 转换写入 ByteArrayOutputStream。总结：WriteObject(Object) 写入 ObjectOutputStream，数据进入 DeflaterOutputStream，然后进入 Base64 编码器，最后进入 ByteArrayOutputStream。映射要么成功，要么异常被捕获并且以运行异常 IllegalStateException 结束。如果数据在 ByteArrayOutputStream 中，则 toString(...) 返回结果。生成的字符串由纯 ASCII 字符组成，也可以使用 US_ASCII 编码。

反向步骤将一个字符串变成一个对象：

```
public static Object deserializeObjectFromBase64( String string ) {
  final byte[] bytes = string.getBytes( StandardCharsets.US_ASCII );

  try ( ByteArrayInputStream bis = new ByteArrayInputStream( bytes );
        InputStream b64is       = Base64.getDecoder().wrap( bis );
        InputStream iis         = new InflaterInputStream( b64is );
        ObjectInputStream ois   = new ObjectInputStream( iis ) ) {
    return ois.readObject();
  }
  catch ( IOException | ClassNotFoundException e ) {
    throw new IllegalStateException( e );
  }
}
```
源代码 6.12 com/tutego/exercise/io/ObjectBase64.java

为了使用 ObjectInputStream 类，需要从 String 生成 InputStream。这是一个小问题，因为 Java 并没有提供一种自然的方法将字符串作为 InputStream 的数据源。谷歌 Guava 或 Apache Commons 等开源库有解决方案，例如以 Apache 类 CharSequenceInputStream 的形式。

Java 只提供另一个方向，例如通过 InputStreamReade 把 InputStream 改编成 Reader，而不是让 Reader 显示成 InputStream，但并没有让我们把 Reader 显

示为 InputStream。因此，当一个字符串需要作为 Reader 出现时，即使常用的 StringReader 也对我们没有帮助。

我们所选择的解决方案是将字符串转换为一个 byte 数组。这其实并不令人满意，因为输入要被运行两次，一次是通过转换，而另一次则是通过从流中读取。另外，这在实践中应该没有什么意义，并增加一些对于我们的使用案例没有真正负担的额外内存。

在将字符串转换为 byte 数组后，ByteArrayInputStream 创建了所需的 InputStream。由 ASCII 字符组成的输入流通过 Base64 解码器生成一个 byte 流。InflaterInputStream 解压数据，最后 ObjectInputStream 通过 readObject( ) 重构对象。序列化的流包含要重构的数据类型的标识符。这个数据类型原则上不可能存在于这个 JVM 上，这就是为什么在这种情况下会抛出 ClassNotFoundException。就像一个可能的 IOException，我们捕捉这些检查过的异常，并创建一个未检查的异常。

测试：6.3.2：序列化的前提条件

Inputs 的实例不能被序列化，因为 Inputs 和 Input 没有实现 Serializable 接口。正确的方式如下：

```
class Inputs implements Serializable {
  public static class Input implements Serializable {
    String input;
  }
  public List<Input> inputs = new ArrayList<>();
}
```

序列化机制递归遍历一个对象图，所有元素都必须是可序列化的。在我们的示例中：

1. 序列化 Inputs。这个类可以序列化吗？是的，那就序列化 ArrayList inputs。
2. 序列化 ArrayList。这个类可以序列化吗？是的，然后在内部序列化带有 Input 条目的数组。
3. 序列化 Inputs。这个类可以序列化吗？是的，然后序列化对象变量 String input。
4. 序列化 String。这个类可以序列化吗？是的，然后序列化这些字符串。

基本数据类型是自动可序列化的，其可见性并不影响序列化过程。静态变量不会被序列化，即使它们被声明为 transient。Java 的许多核心类型本身都是可序列化的，例如 String。此外，枚举和数组也是可序列化的。

任务 6.3.3：保存最后的条目

该任务由一系列陈述组成，但在提出建议的解决方案之前，会提出一个问题：当一个数据类型的属性发生变化和一个对象要在数据流上被反序列化时，会发生什么？答案如下：

```
java.io.InvalidClassException: com.tutego.exercise.net.Inputs$Input;
local class incompatible: stream classdesc serialVersionUID =
-8691588030053894297, local class serialVersionUID = 6463495757449665144
```

有一个异常报告不兼容的 serialVersionUID。背景如下：每个类都有一个标识符，即序列版本 UID（Unique Identifier)。这个 UID 要么是静态地在类中固定，要么是动态地被计算。因为自己的 UID 不存在于来自任务的类中，所以序列化器以类似哈希码的方式计算 UID。这发生在读出和写入时。当一个对象被序列化，这个 UID 会被写入数据流。当读入时，反序列化器检查数据流中的 UID 是否是匹配该类的 UID。如果结构被更改，例如数据类型被更改，则动态 UID 也会更改。这就是显示异常。这两个值分别代表来自数据流的 UID 和计算出来的改变类的 UID。

如果结构上的改变不会导致异常，则必须在代码中手动设置 UID。我们带来了建议的解决方案。

```
class Inputs implements Serializable {
  private static final long serialVersionUID = 1;

  public static class Input implements Serializable {
    private static final long serialVersionUID = 1;
    CharSequence input;
    LocalDateTime localDateTime = LocalDateTime.now();
  }

  public List<Input> inputs = new ArrayList<>();
}
```
源代码 6.13 com/tutego/exercise/io/InputHistory.java

Inputs 和嵌套类 Input 都包含私有静态字段 serialVersionUID；它的初始化的值并不重要。JDK 附带的工具 serialver 会生成与未定义 serialVersionUID 时写入数据流的 UID 完全相同的值，如果类中定义了 serialVersionUID，并且数据流中的 UID 与类的 UID 一致，则反序列化更加宽松：数据流中未知的属性会被忽略，而数据类型已经改变的属性也被跳过。

CharSequence 是一个接口，而接口通常不扩展 Serializable。由于类型检查是在运行时进行的，并且 String 实现了 Serializable 接口，所以没有错误。然而，String 和 CharSequence 会导致不同的 UID。

主程序被嵌入一个 InputHistory 类。该类的构造函数读取一个文件并反序列化输入。另一种方法 addAndSave(...) 更新输入并将结果序列化到文件中。main(...) 方法将所有内容联系在一起。

```java
public class InputHistory {

  private final static Path FILENAME =
      Paths.get( System.getProperty( "java.io.tmpdir" ),
                 "InputHistory.ser" );

  private Inputs inputs;

  InputHistory() {
    try ( InputStream       is = Files.newInputStream( FILENAME );
          ObjectInputStream ois = new ObjectInputStream( is ) ) {
      inputs = (Inputs) ois.readObject();
      inputs.inputs.forEach( input -> System.out.println( input.input ) );
    }
    catch ( IOException | ClassNotFoundException e ) {
      inputs = new Inputs();
      e.printStackTrace();
    }
  }

  void addAndSave( String string ) {
    Inputs.Input newInput = new Inputs.Input();
    newInput.input = string;
    inputs.inputs.add( newInput );

    try ( OutputStream  os = Files.newOutputStream( FILENAME );
```

```java
      ObjectOutputStream oos = new ObjectOutputStream( os ) ) {
    oos.writeObject( inputs );
  }
  catch ( IOException e ) {
    e.printStackTrace();
  }
}

public static void main( String[] args ) {
  InputHistory inputHistory = new InputHistory();
  for ( String line;
      (line = new Scanner( System.in ).nextLine()) != null; ) {
    inputHistory.addAndSave( line );
    System.out.println( line.toUpperCase() );
  }
}
}
```

源代码 6.14 com/tutego/exercise/io/InputHistory.java

该类有两个对象变量：文件名和参考 inputs 的输入。构造函数为文件打开一个 InputStream 并初始化 ObjectInputStream。readObject(...) 开始反序列化，并且如果有异常，则将其捕获并构造一个新的 Inputs 对象。如果可以重建 Inputs，则所有字符串都通过列表的 forEach(...) 方法输出。

addAndSave(String) 方法创建一个新的 Input 对象，将传递的字符串放入该对象，并将新的 Input 对象添加到 inputs 列表中，然后列表通过 ObjectOutputStream 被序列化。除非文件系统有问题，否则不应该出现错误。

main(...) 方法创建一个 InputHistory 对象，它会激活反序列化文件的构造函数。在第一次启动时，这个文件不存在，并抛出一个异常，但是程序不会中断。然后，文件通过控制台输入和保存来创建和增长。下次启动程序时，反序列化应该开始工作，最后输入的字符串应该出现在屏幕上。

第 7 章
网络编程

今天，人们对网络的访问和对本地文件系统的访问一样普遍。

从 Java 1.0 开始，Java 提供了用于开发客户端 – 服务器应用程序的一个网络 API。该 Java 库可以建立加密连接，还提供对超文本传输协议（HTTP）的支持。本章的任务是从网络服务器获取资源并且开发一个具有自己协议的小型客户服务器应用程序。

本章使用的数据类型如下：

- java.net.URL (https://docs.oracle.com/en/java/javase/11/docs/api/java.base/java/net/URL.html)
- java.net.Socket (https://docs.oracle.com/en/java/javase/11/docs/api/java.base/java/net/Socket.html)
- java.net.ServerSocket (https://docs.oracle.com/en/java/javase/11/docs/api/java.base/java/net/ServerSocket.html)
- javax.net.SocketFactory (https://docs.oracle.com/en/java/javase/11/docs/api/java.base/javax/net/SocketFactory.html)
- javax.net.ServerSocketFactory (https://docs.oracle.com/en/java/javase/11/docs/api/java.base/javax/net/ServerSocketFactory.html)
- java.net.DatagramSocket (https://docs.oracle.com/en/java/javase/11/docs/api/java.base/java/net/DatagramSocket.html)

7.1 URL 和 URL 连接

在 Java 中，类 ULR 表示一个 ULR，而 URI 表示一个 ULI。通过 URL 可以打开一个 HTTP 链接，在内部这是由调解器 URLConnection 处理的。

这两个类都不适用于现代 HTTP 调用。只有在 Java 11 中，才有了一个以 HttpClient 为中心的新包 java.net.http。Java 企业框架有其他的解决方案，而且在开源领域也有许多替代方案：

- Jakarta EE 中的客户端 API；
- Webflux 中的 WebClient；
- OkHttp；
- Apache HttpClient；
- 伪装和改造。

7.1.1 通过 URL 下载远程图像

URL 类提供了一个返回 InputStream 的方法，这样就可以读取资源的字节数。
CiaoCiao 船长喜欢在 shiphub.com 上欣赏造型优美的船只图片来放松身心。他想在他的存储器中放一些照片，这样在长途旅行中就有了梦想。

任务：
- 编写一个可以下载指定 URL 资源并存储在本地数据系统中的程序。
- 文件名应以 URL 为基础。

7.1.2 通过 URL 读取远程文本文件 ★

约翰霍普金斯大学系统科学与工程中心 (CSSE) 每天都会在 https://github.com/CSSEGISandData/COVID-19/tree/master/csse_covid_19_data/csse_covid_19_daily_reports 上发布有关全球新冠病毒感染数据的 CSV 文件。2021 年 3 月 1 日的 CSV 文件为 https://raw.githubusercontent.com/CSSEGISandData/COVID-19/master/csse_covid_19_data/csse_covid_19_daily_reports/03-01-2021.csv。

数据开始于：

FIPS,Admin2,Province_State,Country_Region,Last_Update,Lat, Long_,Confirmed, Deaths,Recovered,Active,Combined_Key,Incident_Rate,Case_Fatality_Ratio
,,,Afghanistan,2021-03-02 05:23:30,33.93911, 67.709953, 55733,2444, 49344,3945, Afghanistan,
143.16818690013017,4.385193691349829
,,,Albania,2021-03-02 05:23:30,41.1533,20.1683,107931,1816,70413, 35702, Albania,3750. 4691083466532,
1.6825564481011017

```
,,,Algeria,2021-03-02 05:23:30,28.0339,1.6596,113255,2987,78234,32034,
Algeria,258.2720780438449,
    2.6374111518255265
```

任务：
创建一个新的类 CoronaData，并在其中添加一个新的方法 String findByDate
And-SearchTerm(LocalDate date, String search)。

- findByDateAndSearchTerm(...) 应该创建一个带有日期的 URL 对象。注意文件名采用月、日、年的顺序。
- 打开一个数据流生成的 URL，逐行读取数据流，并过滤掉所有不包含通过的子串 search 的行。
- 最后，一个包含所有带有搜索词的行的字符串被创建，这是 CSV 解析器所不需要的。

举例：

- 调用 findByDateAndSearchTerm(LocalDate.now( ).minusDays(1), "Miguel")，返回从远程 CSV 文件中含有"Miguel"的昨天所有的新冠病毒感染数据。
- 抽样返回：

```
8113,San Miguel,Colorado,US,2020-10-22 04:24:27,38.00450883,
-108.4020725,
100,0,0,100,"San Miguel, Colorado, US",1222.643354933366,0.035047,
San Miguel,New Mexico,US,2020-10-22 04:24:27,35.48014807,
-104.8163562, 151,0,0,151,"San Miguel, New Mexico, US",
553.5799391428677,0.0
```

> **重点：**
> 如果新冠病毒感染数量减小，CSSE 很可能不会发布任何新的文件。旧文件应该被保留。

7.2 HTTP 客户端（Java 11）

原则上可以使用 URLConnection 发出 HTTP 请求，可以更改 HTTP 方法（GET，POST，PUT ……）并设置标头，但这并不方便。因此，在 Java 11 中加入了 HTTP

客户端。它有了新的 API，HTTP 资源可以通过网络更容易地被获取。此外，HTTP 客户端支持 HTTP 1.1 和 HTTP 2 以及同步和异步编程模型。有了这个 API，我们想解决下面的问题；如果不能使用 Java 11，则可以使用 https://github.com/AsyncHttpClient/async-http-client，这是一款适用于 Java 8 的类似库。下一章中，我们将在任务中再次回到 HTTP 客户端，因为 JSON 或 XML 经常被交换。

7.2.1 来自 Hacker News 的头条

Hacker News 是一个专门讨论当前技术趋势的网站。其中的帖子可以通过网络服务访问，文档可以在 https://github.com/HackerNews/API 找到。两个端点如下：

- https://hacker-news.firebaseio.com/v0/topstories.json：返回一个 JSON 数组，其中包含讨论最多的文章的 ID。
- https://hacker-news.firebaseio.com/v0/item/24857356.json：返回带有 ID 为 24857356 消息的 JSON 对象。

任务：
- 创建一个新的类 HackerNews。
- 运行一个新的方法 long[] hackerNewsTopStories( )。
  - 该方法可以构建一个连接到端点 https://hacker-news.firebaseio.com/v0/topstories.json 的 HTTP 客户端；
  - 将 JSON 文件中的返回值分解为 long 值；
  - 所有的 ID 都作为 long[] 返回；
  - 如果出现输入/输出错误，则将返回一个空数组。
- 执行一个新的方法 String news(long id)，它将完整的 JSON 文件作为一个字符串返回。

举例：
带有两个方法的类可以这样使用：

```
System.out.println( Arrays.toString( hackerNewsTopStories() ) );
String newsInJson = news( 24857356 );
System.out.println( newsInJson );
```

7.3 套接字和服务器套接字

操作系统为 TCP/UDP 通信提供了套接字，它可以通过以下类使用：

- java.net.Socket und java.net.ServerSocket（对于 TCP 使用）；
- java.net.DatagramSocket（对于 UDP 使用）。

套接字和服务器套接字的对象是使用构造函数创建的，甚至更好的是使用 javax.net.SocketFactory 和 javax.net.ServerSocketFactory 工厂创建的；对于 UDP 没有工厂。

7.3.1 运行一个演讲服务器和它的客户端

Bonny Brain 很快将参加下一次演讲比赛。她想做充分的准备。她想要一个服务器来运行管理演讲词的应用程序。客户可以连接到该服务器并可以搜索演讲词。

任务：
- 编写一个服务器和客户端。
- 自定义服务器以接收多个连接；访问线程池。
- 线程池应使用最大数量的线程以包含拒绝服务攻击 (DOS)。如果同时连接的最大数量用尽，则客户端必须等待，直到连接再次空闲。

举例：
启动服务器和客户端后，互动可能看起来像这样：

```
sir
You, sir, are an oxygen thief! an
You, sir, are an oxygen thief!
Stop trying to be a smart ass, you're just an ass.
```

7.3.2 实现一个端口的扫描器 ★★

Bonny Brain 安装了新的 Ay! 操作系统，但缺少重要的分析工具。她需要一个工具来检测和报告占用的 TCP/UDP 端口。

任务：
- 编写一个程序，尝试在 0~49 151 的所有 TCP/UDP 端口上注册一个 ServerSocket 和 DatagramSocket。如果成功，则该端口是空闲的，否则端口被占用。
- 在控制台显示被占用的端口，对于已知的端口，显示占用此端口的常用服务的描述。

举例：

这个任务见表 7.1。

表 7.1 任务

Protocol	Port	Service
TCP	135	EPMAP
UPD	137	NetBIOS Name Service
UPD	138	NetBIOS Datagram Service
TCP	139	NetBIOS Session Service
TCP	445	Microsoft-DS Active Directory
TCP	843	Adobe Flash
UPD	1900	Simple Service Discovery Protocol (SSDP)
UPD	3702	Web Services Dynamic Discovery
TCP	5040	
UPD	5050	
UPD	5353	Multicast DNS
UPD	5355	Link-Local Multicast Name Resolution (LLMNR)
TCP	5939	
TCP	6463	
TCP	6942	
TCP	17500	Dropbox
UPD	17500	Dropbox
TCP	17600	
TCP	27017	MongoDB
UPD	42420	

> **信息：**
> 网络端口（engl. network interface）通过计算机网络连接计算机。在下文中，我们总是假设一个 TCP/UDP 端口。网络端口不必是物理硬件的，也可以用软件实现，例如带有 IP 127.0.0.1 (IPv4) 或 ::1 (IPv6) 的环回端口。典型的网络端口也存在于局域网或无线局域网中。操作系统工具可以显示所有网络端口，如 Windows 下的 ipconfig /all 或 Linux 下的 ip a。在 Java 中，网络端口可以通过 java.net.NetworkInterface 查询。每个网络端口有自己的 IP 地址。
> 有两种方法来注册一个 Server-Socket，要么该服务只接收来自特定本地 InetAddress 的请求，或者它接收所有来自本地地址的请求。因此，原则上可以在一块网卡上多次绑定同一个端口，因为在一块网卡上可以配置任意数量的网络端口，因为它们拥有不同的 IP 地址。
> 为了解决这个问题，可以做一个简单的测试，在所有的网络端口注册 Socket-Socket。如果失败，就表示一个网络端口处于活动状态。对于我们来说这可以作为标准，表示该服务在任意网络端口被占用。

7.4 建议解决方案

任务 7.1.1：通过 URL 下载远程图像

```java
public static void downloadImage( URL url ) throws IOException {
  try ( InputStream inputStream = url.openStream() ) {
    String filename = url.toString().replaceAll( "[^a-zA-Z0-9_.-]", "_" );
    Files.copy( inputStream, Paths.get( filename ),
      StandardCopyOption.REPLACE_EXISTING );
  }
}
```

源代码 7.1 com/tutego/exercise/net/ImageDownloader.java

该解决方案分成以下 3 步：

1. URL 对象通过 openStream() 方法提供一个指向资源字节的 InputStream。由于打开的流应该及时关闭，所以我们将打开流的操作放在一个 try-with-resources 块中。
2. 目标文件的名称来源于 URL，但在处理文件名时需要小心，因为 URL 中的某些字符并不总是有效的文件名字符。维基百科页面 https://en.wikipedia.org/wiki/Filename#Comparison_of_filename_limitations 总结了一些文件系统中特殊字符的限制。同样，在 URL 中也不允许使用所有字符，甚至简单的路径分隔符 "/" 也会引发问题。因此，我们需要替换所有可能引起问题的字符。解决方案选择规范的表达式 [^a-zA-Z0-9_.-] 将所有不是字母、数字、"_"、"."或 "-" 的字符替换为下划线。这为每个文件系统提供了一个安全的文件名。但是，URL 不是唯一的，例如 https://www.penisland.net/?; 和 https://www.penisland.net/?, 都会变成 http___www.penisland.net___。

3. 第三步是使用 Files 类的 copy(...) 方法。有两个方法：一个用于从文件中读取数据并写入 OutputStream, 另一个用于从 InputStream 中读取所有数据并写入文件。我们使用第二个方法。所有字节都从 InputStream 中读取，并且写入新的目标文件。copy(...) 的第三个参数是可变参数并代表属性：StandardCopyOption.REPLACE_EXISTING 表示现有文件将被覆盖，否则现有文件将出现异常。

任务 7.1.2: 通过 URL 读取远程文本文件

```java
private static final String URL_TEMPLATE =
    "https://raw.githubusercontent.com/CSSEGISandData/COVID-19/master/csse_covid_19_data/csse_covid_19_daily_reports/%s.csv";
public static String findByDateAndSearchTerm( LocalDate date, String search ) {
  String url = String.format( URL_TEMPLATE,
    date.format( DateTimeFormatter.ofPattern( "MM-dd-yyyy" ) ) );

  try ( InputStream is = new URL( url ).openStream();
        Reader isr = new InputStreamReader( is, StandardCharsets.UTF_8 );
        BufferedReader br = new BufferedReader( isr ) ) {
    return br.lines()
        .filter( line -> line.contains( search ) )
        .collect( Collectors.joining( "\n" ) );
  }
  catch ( MalformedURLException e ) {
    System.err.println( "Malformed URL format of " + url );
  }
  catch ( IOException e ) {
    e.printStackTrace();
```

```
    }
    return "";
}
```

源代码 7.2 com/tutego/exercise/net/CoronaData.java

第一步是建立 URL。URL 会包含日期，我们将其作为一个参数接收。通常不能直接在 LocalDate 对象上调用 toString( ) 方法，因为它返回一个依据 ISO 8601-Notation 的字符串；CSSE 使用其他顺序表示指定日期段。该程序使用 DateTimeFormatter 构建自己的日期格式，其顺序先是月，然后是日，最后是年。

从这个动态构造的字符串创立一个 URL 对象，并调用中央方法 openStream( )，该方法返回一个 InputStream。由于要逐个读取字符，所以我们把这个二进制的 InputStream 转换成一个 Reader。BufferedReader 非常适合逐行读取，返回 Stream<String> 的 lines( ) 方法非常适用。通过一个过滤器，我们在流中留下包含搜索字符串的行，最后将所有行串联成一个大字符串并返回。

有两种可能的例外情况。其中一种情况是 URL 格式错误，在这种情况下会出现 MalformedURLException，或者在读取过程中出现连接错误。我们拦截这些异常，将信息输出显示在屏幕上，并返回一个空字符串，因为即使在出错的情况下也没有找到任何东西。

任务 7.2.1 来自 Hacker News 的头条

```
private static final HttpClient client = HttpClient.newHttpClient();

public static long[] hackerNewsTopStories() {
  HttpRequest request = HttpRequest
      .newBuilder( URI.create(
      "https://hacker-news.firebaseio.com/v0/topstories.json" ) )
      .timeout( Duration.ofSeconds( 5 ) )
      .build();

  try {
    HttpResponse<InputStream> response =
        client.send( request, HttpResponse.BodyHandlers.ofInputStream() );
    Scanner scanner = new Scanner( response.body() ).useDelimiter(
      "[,\\[\\]]" );
    return scanner.tokens().mapToLong( Long::parseLong ).toArray();
```

```
  }
  catch ( IOException | InterruptedException e ) {
    e.printStackTrace();
  }
  return new long[ 0 ];
}
```

源代码 7.3 com/tutego/exercise/net/HackerNews.java

```
public static String news( long id ) {
  HttpRequest request = HttpRequest
    .newBuilder( URI.create( "https://hacker-news.firebaseio.com/v0/item/" +
    id + ".json" ) )
    .timeout( Duration.ofSeconds( 5 ) )
    .build();

  try {
    return client.send( request,
      HttpResponse.BodyHandlers.ofString() ).body();
  }
  catch ( IOException | InterruptedException e ) {
    e.printStackTrace();
    return "";
  }
}
```

源代码 7.4 com/tutego/exercise/net/HackerNews.java

对于我们的例子，创建一个 HttpClient 作为类变量。由于两个方法都是静态的，所以在现实中，对象的预先配置会被提前准备。如果对象方法正在访问 HttpClient，并且来自多个线程，那么每个线程都应该使用自己的 HttpClient 对象。

这个 send(…) 方法有两个参数：

```
<T> HttpResponse<T> send(HttpRequest request,
                HttpResponse.BodyHandler<T> responseBodyHandler)
    throws IOException, InterruptedException;
```

先创建并配置一个可以单独被访问的 HttpRequest。这可以通过 HttpRequest.

newBuilder( ) 和 HttpRequest.newBuilder(URI) 来完成。第二个版本保存了后来对 uri(...) 的调用。超时是可选的，由程序设定为 5 秒。最后的 build( ) 方法返回一个由 HttpClient 执行的 HttpRequest 对象。API 的原理就是：一旦创建和配置了 HttpClient，就会运行不同的 HttpRequest 调用。

send(...) 方法的第二个参数决定了如何获取结果。对于下面的 Scanner，在实际中 InputStream 会调用 HttpRe- sponse.BodyHandlers.ofInputStream( )。我们使用 body( ) 从 HttpResponse<InputStream>response 中读取正文，因此正文一开始不是数据，而只是该数据的 InputStream。InputStream 设置 Scanner。在我们的例子中，编码不是必需的，因为 JSON 文件由纯 ASCII 字符组成。然而，Scanner 设置了一个分界线，将 [和] 视为分隔符。自 Java 9 以来，Scanner 提供了一个有用的方法 tokens( )，它给了我们一个所有标记的流，这在我们的例子中是数字。mapToLong(...) 转换每个文本表示法的数字进入 LongStream，toArray( ) 将流的所有元素作为数组返回。如果有错误，它们会被捕获并报告，然后返回一个空数组。

另一种方法就是使用 news(long id)，让 HttpRequest 和一个信息的指定 ID 关联。执行比较容易，因为我们不需要解析输出，只要通过 HttpResponse.BodyHandlers.ofString( ) 来确定返回一个 String 作为结果。

来自 Web 服务的 JSON 文件通常会被转换为 Java 对象。

下一章将介绍如何做到这一点。

任务 7.3.1：运行一个演讲服务器和它的客户端

服务器的任务是对传入的连接作出反应，并挑出所有带辱骂内容的字符串。

```
public class SlangingMatchServer {

  private static final int PORT              = 10_000;
  private static final int MAXIMUM_POOL_SIZE = 10_000;

  public static void main( String[] args ) throws IOException {
    Executor executor = new ThreadPoolExecutor(
      0, MAXIMUM_POOL_SIZE,
      60, TimeUnit.SECONDS,
      new SynchronousQueue<>() );

    try ( ServerSocket serverSocket =
        ServerSocketFactory.getDefault().createServerSocket( PORT ) ) {
      System.out.println( "Server running at port " +
```

```java
      serverSocket.getLocalPort() );

    while ( Thread.currentThread().isInterrupted() ) {
      Socket socket = serverSocket.accept();
      executor.execute( () -> handleConnection( socket ) );
    }
  }
}

private static void handleConnection( Socket socket ) {
  try ( Socket __ = socket; // try ( socket ) since Java 9
        Scanner requestReader = new Scanner( socket.getInputStream(),
          StandardCharsets.UTF_8 );
        PrintWriter responseWriter = new PrintWriter(
          socket.getOutputStream(),
          true, StandardCharsets.UTF_8 ) ) {
    String request = requestReader.nextLine();
    responseWriter.println( searchInsult( request ) );
  }
  catch ( IOException e ) {
    e.printStackTrace();
  }
}

private static String searchInsult( String search ) {
  return Stream.of( "You, sir, are an oxygen thief!",
    "Stop trying to be a smart ass, you're just an ass.",
    "Shock me, say something intelligent." )
    .filter( s -> s.toLowerCase().contains( search.toLowerCase() ) )
    .collect( Collectors.joining( "\n" ) );
  }
}
```

源代码 7.5 com/tutego/exercise/net/SlangingMatchServer.java

main(...) 方法为预定义的端口 10000 准备一个服务器套接字，进入一个原则上可以通过中断结束的无限循环，并等待传入的连接。在 accept() 上有线程阻塞，并且阻塞仅在传入客户端时清除。在这种情况下，通过线程池调用自己的 handleCollection(..) 方法。Thread-PoolExecutor 是线程池，到现在为止，我们总是可以把它设置为 Executors.new-CachedThreadPool()，但此时并发连接的数量是无限的。并发线程数可以通过 ThreadPoolExecutor 构造函数来指定。

当 accept() 返回时，返回的是客户端套接字，它会成为 handleConnection(...) 的参数。我们将 handleConnection(...) 的调用移至一个 Runnable，它被传递给 Executor，从而在后台执行。当线程在后台接收客户端服务器的通信时。无限循环返回 accept()，以便能够快速为下一个感兴趣方提供服务。这时不能关闭 Socket，也不允许如下操作：

```
try ( Socket socket = serverSocket.accept() ) {
  executor.execute( () -> handleConnection( socket ) );
}
```

处理过程是异步的。try-mit-Ressourcen 块的 close() 会直接通过 execute(…) 发送，否则当 handleConnection(…) 刚开始通信时，Socket 会快速关闭。

关注一下 handleConnection(…)：这个 Socket 是 AutoCloseable 的，try-mit-Ressourcen 块最后会关闭这个 Socket。这里有一个不寻常的例子，实际上无须声明一个新的资源变量，但缩写符号从 Java 9 才开始被允许使用。

在通信中，InputStream 和 OutputStream 是必要的。这些是需要通过 try-mit-Ressourcen 块来关闭的资源。该协议允许通过网络发送字符串，因此，InputStream 和 OutputStream 被升级为基于面向字符的类型。Scanner 可以在构造函数构建一个 InputStream 和一个编码，然后可以使用 nextLine() 读取一行。我们将输出写入 PrintWriter；在这里，构造函数也可以接收 OutputStream 并指定编码。第二个参数 true 是很重要的，因为它控制在行末字符刷新缓冲区。println(...) 将结果写入 PrintWriter，换行表示缓冲区刷新。使用 catch 块结束 try-mit-Ressourcen 块，所有资源都被关闭，Socket 作为本地资源返回给操作系统。

应用方法 searchInsult(...) 检查搜索词是否包含在给定的字符串中，并将所有结果用换行符连接起来。

客户端有一个类似的逻辑，它不需要监听传入的连接，而是创建一个连接。

```
public class SlangingMatchClient {

  private static final String HOST = "localhost";
  private static final int    PORT = 10_000;
```

```java
public static void main( String[] args ) throws IOException {
  while ( true ) {
    String request = new Scanner( System.in ).nextLine();
    remoteSearchInsult( request );
  }
}

private static void remoteSearchInsult( String search ) throws IOException {
  try ( Socket socket = SocketFactory.getDefault().createSocket( HOST, PORT );
       PrintWriter requestWriter =
           new PrintWriter( socket.getOutputStream(), true,
             StandardCharsets.UTF_8 );
       BufferedReader responseReader = new BufferedReader(
           new InputStreamReader( socket.getInputStream(),
             StandardCharsets.UTF_8 ) ) ) {
    requestWriter.println( search );
    System.out.println( responseReader.lines().collect(
      Collectors.joining( "\n" ) ) );
  }
}
```

源代码 7.6: com/tutego/exercise/net/SlangingMatchClient.java

main(...) 方法包含一个无限循环，向用户询问一个字符串并将其传递给 remoteSearchInsult(String)。新方法负责与服务器通信。

套接字工厂为 localhost 和所需端口提供一个 Socket 对象。套接字是最终返回操作系统的本机资源。像往常一样通过 try-with-resources 块完成关闭，该块也关闭 InputStream 和 OutputStream。

PrintWriter 再次写入字符串。客户端使用 BufferedReader 进行读取，这样的优点是通过 lines( ) 方法可以返回一个 Stream 行。读取的行将和 Collector 连接并输出。不能直接把一个 Reader 和一个 InputStream 连接，因此需要在 InputStream 上启用 Reader API 的 InputStreamReader 装饰器。

任务 7.3.2：实现一个端口的扫描器

PortScanner 类使用它自己的枚举 Protocol，包括常量 TCP 和 UDP、一个公共对象方法 isAvailable(int)、私人辅助方法 openSocket(int) 和静态方法 serviceName(int)。

```java
    enum Protocol {
        TCP {
            @Override AutoCloseable openSocket( int port ) throws IOException {
                return ServerSocketFactory.getDefault().createServerSocket( port );
            }
        },
        UDP {
            @Override AutoCloseable openSocket( int port ) throws IOException {
                return new DatagramSocket( port );
            }
        };

        abstract AutoCloseable openSocket( int port ) throws IOException;

        public boolean isAvailable( int port ) {
            try ( AutoCloseable _ = openSocket( port ) ) { return true; }
            catch ( Exception e ) { return false; }
        }

    private static final String COMPRESSED_SERVICE_NAMES =
            "7 Echo\n13 Daytime\n20 FTP\n21 FTP\n22 SSH\n23 Telnet\n25 SMTP\n53 DNS\n80 HTTP\n"
          + "135 EPMAP\n137 NetBIOS Name Service\n138 NetBIOS Datagram  Service\n139 NetBIOS Session Service\n"
          + "445 Microsoft-DS Active Directory\n843 Adobe Flash\n"
          + "1900 Simple Service Discovery Protocol (SSDP)\n3702 Web Services Dynamic Discovery\n"
          + "5353 Multicast DNS\n5355 Link-Local Multicast Name Resolution (LLMNR)\n"
          + "17500 Dropbox\n27017 MongoDB\n";

    private static final Map<Integer, String> SERVICE_NAMES =
         Pattern.compile( "\n" )
                .splitAsStream( COMPRESSED_SERVICE_NAMES )
                .map( Scanner::new )
                .collect( Collectors.toMap( Scanner::nextInt, Scanner::nextLine ) );
```

```
    static String serviceName( int port ) {
      return Optional.ofNullable( SERVICE_NAMES.get( port ) ).orElse( "" );
    }
  }
```

源代码 7.7 com/tutego/exercise/net/PortScanner.java

枚举类型的 Protocol 声明了一个抽象方法 openSocket(int) 用于打开连接，因为 TCP 和 UDP 的代码不同；两个枚举元素相应地实现了抽象方法。在 TCP 的情况下，应用程序通过 ServerSocketFactory 构建一个 ServerSocket，或者通过 DatagramSocket 的构造函数构建一个 DatagramSocket。虽然 ServerSocket 和 DatagramSocket 是不同的类型，但两者都实现了 AutoCloseable 接口，而且这种类型也返回 openSocket(int)，因为只有这种类型与 isAvailable(int) 有关。

isAvailable(int) 的任务是确定该端口是否已经被占用。为此，它调用 openSocket(...)，如果没有异常，则说明该端口是空闲的，连接可以再次被立刻关闭，处理在 AutoCloseable 上的 try-mit-Ressourcen serviceName(int) 访问先前构造的 Map。在类内部有一个常量 COMPRESSED_SERVICE_NAMES，它可以很容易地来自一个文件。该字符串包含端口号并以空格分隔一个简短的描述，然后以换行符结束。从 Java 15 开始，可以很好地使用文本块，从而可以在代码中优雅地格式化字符串。Stream 表达式通过将 String 分解为行来准备 Map 这个解决方案与 Java 8 兼容。从 Java 11 开始，String 方法 lines( ) 更简单地解决了这个问题。Stream 由传输到 Scanner 对象的构造函数的行组成，因此 nextInt( ) 返回关联内存的键，nextLine( ) 返回与端口号相关的简短描述。serviceName(int) 可以访问这个关联内存 SERVICE_NAMES，如果没有端口号的描述，则返回一个空字符串。在另一个模型中，当然可以直接返回 Optional.empty( )，但这里的空字符串很方便。

main(...) 方法使用 PortScanner 方法：

```
    final int MIN_SYSTEM_PORT       =      0;
    // final int MAX_SYSTEM_PORT       = 1023;
    // final int MIN_REGISTERED_PORT = 1024; final int MAX_REGISTERED_
PORT = 49151;

    System.out.println( "Protocol  Port    Service" );
    for ( int port = MIN_SYSTEM_PORT; port <= MAX_REGISTERED_PORT; port++ ) {
      for ( Protocol protocol : Protocol.values() )
        if ( ! protocol.isAvailable( port ) )
          System.out.printf( "%s   %5d    %s%n",
```

```
                    protocol, port, Protocol.serviceName( port ) );
}
```
源代码 7.8: com/tutego/exercise/net/PortScanner.java

首先定义了 4 个常数。它们代表了我们的端口扫描器所要运行的端口范围的极限。

在代码示例中，用两个常数表示上限和下限，因为在 MAX_SYSTEM_PORT 之后继续使用 MIN_REGISTERED_PORT。通过 for 循环来使用所有注册的 0~4 915 的端口号。Protocol.values( ) 返回一个包含两个枚举元素的数组 TCP 和 UDP，如果在枚举上声明的方法 isAvailable(...) 显示一个阻塞的端口，就会在打印控制台输出。

第 8 章
用 Java 处理 XML、JSON 等数据格式的文件

交换文件的两种重要数据格式是 XML 和 JSON。XML 是历史上较早的数据类型，现在经常在服务器和 JavaScript 应用程序之间的通信中使用 JSON。JSON 文件也经常被用作配置文件。

虽然 Java SE 为读写 XML 文件提供了不同的类，但 JSON 支持只在 Java 企业版中或通过补充的开源库提供。因此，本章中的许多任务使用了外部库。

描述语言构成了文件格式的一个重要类别。它们定义了数据的结构。最重要的格式包括 HTML、XML、JSON 和 PDF。

除了支持属性文件和可以处理 ZIP 档案外，Java 不支持其他数据格式。这对 CSV 文件、PDF 或 Office 文件来说尤其如此。幸运的是，有几十个开源库填补了这一空白，因此人们不必自己编程实现这个功能。

本章使用的数量类型如下：

- javax.xml.stream.XMLOutputFactory (https://docs.oracle.com/en/java/javase/11/docs/api/java.xml/javax/xml/stream/XMLOutputFactory.html)

- javax.xml.stream.XMLStreamWriter (https://docs.oracle.com/en/java/javase/11/docs/api/java.xml/javax/xml/stream/XMLStreamWriter.html)

- javax.xml.stream.XMLStreamException (https://docs.oracle.com/en/java/javase/11/docs/api/java.xml/javax/xml/stream/XMLStreamException.html)

- javax.xml.stream.XMLInputFactory (https://docs.oracle.com/en/java/javase/11/docs/api/java.xml/javax/xml/stream/XMLInputFactory.html)

- javax.xml.stream.XMLStreamReader (https://docs.oracle.com/en/java/javase/11/docs/api/java.xml/javax/xml/stream/XMLStreamReader.html)

- javax.xml.bind.JAXB (https://docs.oracle.com/javaee/7/api/javax/xml/bind/JAXB.html)

- javax.xml.bind.DataBindingException (https://docs.oracle.com/javaee/7/api/javax/xml/bind/DataBindingException.html)
- https://github.com/FasterXML/jackson

8.1 用 Java 处理 XML 文件

处理 XML 文件有不同的 Java API。一种解决方案是将完整的 XML 对象保存在内存中，另一种解决方案是使用数据流。StAX 是一个 Pull-API，通过它元素可以从数据流中主动被获取并被写入。该处理模型非常适合不必完全在内存中存储的大型文件。

JAXB 提供了一种简单的方法来将 Java 对象转换为 XML 文件，并在以后将 XML 转换回 Java 对象。在注释或外部配置文件的帮助下，映射可以得到精确控制。

8.1.1 编写带有配方的 XML 文件

CiaoCiao 船长有很多配方，因此他需要一个数据库。他有几个数据库管理系统的方案，他想看看它们是否能导入他的所有配方。

他自己的配方是 RecipeML 格式，这是一种松散的 XML 格式：http://www.formatdata.com/recipeml/。

在 https://dsquirrel.tripod.com/recipeml/indexrecipes2.html 中有一个大的数据库，一个来自"KeyGourmet"的举例如下：

```
<?xml version="1.0" encoding="UTF-8"?>
<recipeml version="0.5">
  <recipe>
    <head>
      <title>11 Minute Strawberry Jam</title>
      <categories>
        <cat>Canning</cat>
        <cat>Preserves</cat>
        <cat>Jams & jell</cat>
      </categories>
      <yield>8</yield>
    </head>
    <ingredients>
      <ing>
        <amt>
```

```xml
          <qty>3</qty>
          <unit>cups</unit>
        </amt>
        <item>Strawberries</item>
      </ing>
      <ing>
        <amt>
          <qty>3</qty>
          <unit>cups</unit>
        </amt>
        <item>Sugar</item>
      </ing>
    </ingredients>
    <directions>
      <step>Put the strawberries in a pan.</step>
      <step>Add 1 cup of sugar.</step>
      <step>Bring to a boil and boil for 4 minutes.</step>
      <step>Add the second cup of sugar and boil again for 4 minutes.</step>
      <step>Then add the third cup of sugar and boil for 3 minutes.</step>
      <step>Remove from stove, cool, stir occasionally.</step>
      <step>Pour in jars and seal.</step>
    </directions>
  </recipe>
</recipeml>
```

任务：
编写一个以 RecipeML 格式输出 XML 文件的程序。

8.1.2　检查所有图片是否有一个 alt 属性

HTML 文件中的图片应该总带有一个 alt 属性。

任务：

- 执行一个 XHTML 验证器，该验证器报告每个 img 标记是否带有 alt 属性。
- 可以使用诸如 http://tutego.de/download/index.xhtml 作为 XHTML 文件。

8.1.3 编写带有 JAXB 的 Java 对象

JAXB 简化了对 XML 文件的访问，因为它允许从 Java 对象方便映射到 XML 文件，反之亦然。

JAXB 包含在 Java 6 的标准版中，并在 Java 11 中被删除。要为当前的 Java 版本做好准备，应在 POM 文件中包含以下内容：

```xml
<dependency>
  <groupId>javax.xml.bind</groupId>
  <artifactId>jaxb-api</artifactId>
  <version>2.3.1</version>
</dependency>
<dependency>
  <groupId>org.glassfish.jaxb</groupId>
  <artifactId>jaxb-runtime</artifactId>
  <version>2.3.3</version>
  <scope>runtime</scope>
</dependency>
```

任务：

▶ 编写 JAXB-Beans,，以便可以生成以下 XML 文件：

```xml
<?xml version="1.0" encoding="UTF-8" standalone="yes"?>
<ingredients>
    <ing>
        <amt>
            <qty>3</qty>
            <unit>cups</unit>
        </amt>
        <item>Sugar</item>
    </ing>
    <ing>
        <amt>
            <qty>3</qty>
            <unit>cups</unit>
        </amt>
    </ing>
```

</ingredients>
- 创建类 Ingredients, Ing, Amt an。
- 给出对应类的对象变量；如果它们是公共的，那么一切正常。
- 考虑使用哪个注释。

8.1.4　阅读笑话并开心地笑★★

　　Bonny Brain 阅读简单的笑话也会笑，她对于笑话永远不会满足。她在互联网上找到这个一直有最新笑话的网站：https://sv443.net/jokeapi/v2/joke/Any?format=xml。

　　这个格式是适合数据传输的 XML，但我们是 Java 开发人员，希望一切都在对象中！通过 JAXB，XML 文件要被读取并被转换为 Java 对象，以便以后可以开发单独的输出。

　　JAXB-Beans 从 XML 文件中自动生成。笑话页面的代码如下——别担心，你不必理解它。

```
<xs:schema attributeFormDefault="unqualified" elementFormDefault="qualified" xmlns:xs="http://www.w3.org/2001/XMLSchema">
  <xs:element name="data">
    <xs:complexType>
```

```xml
      <xs:sequence>
        <xs:element type="xs:string" name="category" />
        <xs:element type="xs:string" name="type" />
        <xs:element name="flags">
          <xs:complexType>
            <xs:sequence>
              <xs:element type="xs:boolean" name="nsfw" />
              <xs:element type="xs:boolean" name="religious" />
              <xs:element type="xs:boolean" name="political" />
              <xs:element type="xs:boolean" name="racist" />
              <xs:element type="xs:boolean" name="sexist" />
            </xs:sequence>
          </xs:complexType>
        </xs:element>
        <xs:element type="xs:string" name="setup" />
        <xs:element type="xs:string" name="delivery" />
        <xs:element type="xs:int" name="id" />
        <xs:element type="xs:string" name="error" />
      </xs:sequence>
    </xs:complexType>
  </xs:element>
</xs:schema>
```

供应商提供的不是方案，而是通过网页 https://www.freeformat-ter.com/xsd-generator.html 生成的 XML 文件。

任务：

- 在 http://tutego.de/download/jokes.xsd 中加载 XML 模式定义，并且将该文件放在 Maven 目录 /src/main/resources 中。
- 在 POM 文件中添加以下元素：

```xml
<build>
<plugins>
  <plugin>
    <groupId>org.codehaus.mojo</groupId>
    <artifactId>jaxb2-maven-plugin</artifactId>
    <version>2.5.0</version>
```

```xml
      <executions>
        <execution>
          <id>xjc</id>
          <goals>
            <goal>xjc</goal>
          </goals>
        </execution>
      </executions>
      <configuration>
        <packageName>com.tutego.exercise.xml.joke</packageName>
        <sources>
          <source>src/main/resources/jokes.xsd</source>
        </sources>
        <generateEpisode>false</generateEpisode>
        <outputDirectory>${basedir}/src/main/java</outputDirectory>
        <clearOutputDir>false</clearOutputDir>
        <noGeneratedHeaderComments>true</noGeneratedHeaderComments>
        <locale>en</locale>
      </configuration>
    </plugin>
  </plugins>
</build>
```

插件部分集成并配置了 org.codehaus.mojo:jaxb2-maven-plugin；所有选项都在 https://www.mojohaus.org/jaxb2-maven- plugin/Documentation/v2.5.0/index.html 中进行了解释。

- 从命令行启动 mvn generate-sources。在 com.tutego.exercice.xml.joke 包中创建了两个类：
  - Data；
  - ObjectFactory。
- 使用 JAXB 从 URL https://sv443.net/jokeapi/v2/joke/Any?format=xml 获取一个笑话并将其转换为一个对象。

8.2　JSON

　　Java SE 不支持 JSON，但 Jakarta EE 支持。一个常用的方案是 Jackson，它也可以将 Java 对象映射到 JSON 对象中，并从 JSON 对象中重构 Java 对象。Jackson 的模块化程度很高，它包含 3 个核心模块（Streaming, Annotations 和 Databind）以及一些第三方模块，特别对于诸如 Guava, javax.money 的数据类型进行了性能优化，例如生成字节码而不是依赖反射。

Databind 模块对流和注解有依赖性，因此开发者可以通过它获得所有核心功能。在 Maven POM 中包含以下依赖关系：

```
<dependency>
  <groupId>com.fasterxml.jackson.core</groupId>
  <artifactId>jackson-databind</artifactId>
  <version>2.12.3</version>
</dependency>
```

8.2.1 黑客新闻：评估 JSON 文件

黑客新闻（https://news.ycombinator.com）已经在第 7 章"网络编程"中简单介绍了。URL（https://hacker-news.firebaseio.com/v0/item/24857356.json）返回一个 ID 为 24857356 的消息的 JSON 对象。响应看起来是这样的（经过格式化并对内容进行了适当精简，以适合孩子们阅读）：

```
{
    "by":"luu",
    "descendants":257,
    "id":24857356,
    "kids":[
       24858151,
       24857761,
       24858192,
       24858887
    ],
    "score":353,
    "time":1603370419,
    "title":"The physiological effects of slow breathing in the healthy human",
    "type":"story",
    "url":"https://breathe.ersjournals.com/content/13/4/298"
}
```

使用 Jackson，这个 JSON 可以被转换为 Map：

```
ObjectMapper mapper = new ObjectMapper();
Map map = mapper.readValue( src, Map.class );
```

Src 有不同的数据来源，例如类型 String, File, Reader, InputStream, URL …

任务：

编写新的方法 Map<?, ?> news(long id)，它通过 Jakson 的帮助，检索在 unter "https://hacker-news.firebaseio.com/v0/item/" + id + ".json" 中的 JSON 文件，将其转换为 Map 并返回。

举例：

- news(24857356).get("title") → "The physiological effects of slow breathing in the healthy human";
- news(111111).get("title") → null。

8.2.2　以 JSON 格式读取写入编辑器的配置★★

程序员为 CiaoCiao 船长开发了一个新的编辑器，配置将被保存在一个 JSON 文件中。

任务：

- 编写一个 Settings 类，以便可以映射以下配置：
```
{
  "editor" : {
    "cursorStyle" : "line",
    "folding" : true,
    "fontFamily" : [ "Consolas, 'Courier New', monospace" ],
    "fontSize" : 22, "fontWeight" : "normal"
  },
  "workbench" : {
    "colorTheme" : "Default Dark+"
  },
  "terminal" : {
    "integrated.unicodeVersion" : "11"
  }
}
```
- JSON 文件很好地展示了数据类型：cursorStyle ist String, folding ist boolean（fontFamily 是数组或列表）。

- 如果一个属性未被设置，即为 0，则它将不会被写入。
- 对于终端来说，其所包含的键值是未知的，它们将被包含在一个 map<string, string> 中。

8.3 HTML

HTML 是一种重要的标记语言。Java 标准库不提供对 HTML 文件的支持，除了 javax.swing.JEditorPane 可以显示 HTML 3.2 和一个 CSS 1.0 子集。

为了使 Java 程序能够正确、有效地读取写入 HTML 文件和节点，不得不求助（开源的）库。

8.3.1 使用 jsoup 加载维基百科图像 ★★

流行的开源库 jsoup（https://jsoup.org/）加载网页的内容，并在内存中以树状形式表示内容。

在 POM 中包括以下依赖关系：

```
<dependency>
  <groupId>org.jsoup</groupId>
  <artifactId>jsoup</artifactId>
  <version>1.13.1</version>
</dependency>
```

任务：

- 学习以下列子——https://jsoup.org/cookbook/extracting-data/dom-navigation 和 https://jsoup.org/cookbook/extracting-data/selector-syntax。
- 从维基百科主页上查询所有图片并将其保存在自己的文件系统中。

8.4 Office 文件

在文字处理和电子表格方面，微软公司的 Office 软件仍然处于领先地位。多年来，二进制文件格式已经广为人知，并且可以用于 Java 库读写，处理 Office 文件变得更加容易，因为这些文件核心就是 XML 文件，并被组合在一个 ZIP 档案中。Java 对它的支持是非常好的。

8.4.1 生成带截图的 Word 文件

阅读关于 POI 的 Wikipedia 条目：https://de.wikipedia.org/wiki/Apache_POI。

任务：

1. 在 POM 中为 Maven 添加以下内容（包含 Apache POI 和 DOCX 的必要依赖项）：

   ```
   <dependency>
     <groupId>org.apache.poi</groupId>
     <artifactId>poi-ooxml</artifactId>
     <version>5.0.0</version>
   </dependency>
   ```

2. 学习源代码：SimpleImages.java unter http://svn.apache.org/repos/ asf/poi/trunk/poi-examples/src/main/java/org/apache/poi/examples/xwpf/user- model/SimpleImages.java。

3. 可以用 Java 进行截图，像这样：

   ```
   private static byte[] getScreenCapture() throws AWTException, IOException {
     BufferedImage screenCapture = new Robot().createScreenCapture(
       SCREEN_SIZE );
     ByteArrayOutputStream os = new ByteArrayOutputStream();
     ImageIO.write( screenCapture, "jpeg", os );
     return os.toByteArray();
   }
   ```

4. 编写一个 Java 程序，每 2 秒截屏一次，持续 20 秒，并将图像附加到 Word 文件中。

8.5 归档

带有元数据的文件被归档。一个著名和流行的归档格式是 ZIP，它不仅可以汇总数据，还可以压缩数据。许多归档格式还可以以加密的形式保存文件并附带校验码，以便以后传送中的错误可以被检测到。

Java 为压缩提供了两种可能：从 Java 7 开始，有一个 ZIP 文件系统提供者；从 Java 1.0 开始，有了 ZipFile 类和 ZipEntry 类。

8.5.1 播放 ZIP 档案中的昆虫声音 ★★

Bonny Brain 喜欢听昆虫的声音，他使用 WAV 格式收集来自 https://catalog.data.gov/dataset/bug-bytes-sound-library-stored-product-insect-pest-sounds 的音频，那里的各种音频文件以 ZIP 档案的形式提供下载。

任务：

▶ 从 https://christian-schlichtherle.bitbucket.io/true- zip/truezip-path/ 中学习文档。

- 在 Maven POM 中包含两个依赖项：
  ```
  <dependency>
    <groupId>de.schlichtherle.truezip</groupId>
    <artifactId>truezip-path</artifactId>
    <version>7.7.10</version>
  </dependency>

  <dependency>
    <groupId>de.schlichtherle.truezip</groupId>
    <artifactId>truezip-driver-zip</artifactId>
    <version>7.7.10</version>
  </dependency>
  ```
- 下载带有昆虫声音的 ZIP 档案，但不要解压。
- 为 ZIP 档案创建一个 TPath 对象。
- 将 ZIP 档案中的所有文件名转移到一个列表中，Files.new- DirectoryStream(…) 会提供帮助。
  - 编写一个无限循环。
  - 选择一个随机的 WAV 文件。

  用 Files.newInputStream(…) 打开随机选择的 WAV 文件，用 BufferedInputStream 装饰它，并打开 AudioSystem.getAudioInput-Stream(…)。播放 WAV 文件，并访问以下代码，其中 ais 是 AudioInputStream：

  ```
  Clip clip = AudioSystem.getClip();
  clip.open( ais );
  clip.start();
  TimeUnit.MICROSECONDS.sleep( clip.getMicrosecondLength() + 50 );
  clip.close();
  ```

  在《Ciao Ciao 船长征服 Java：喂养大脑更好的 Java 技能练习卷 I》第 8 章中已经使用了 javax.sound API。

8.6 建议解决方案

任务 8.1.1：编写带有配方的 XML 文件

建议解决方案虽然很长但不复杂。它从两种类型开始：

```
class Recipe {
  String head$title;
```

```java
    List<String> head$categories;
    String head$yield;
    List<Ingredient> ingredients;

    List<String> directions;

    Recipe( String head$title, List<String> head$categories, String head$yield,
           List<Ingredient> ingredients, List<String> directions ) {
      this.head$title = head$title;
      this.head$categories = head$categories;
      this.head$yield = head$yield;
      this.ingredients = ingredients;
      this.directions = directions;
    }

    static class Ingredient {
      String ing$amt$qty;
      String ing$amt$unit;
      String ing$item;

      Ingredient( String ing$amt$qty, String ing$amt$unit, String ing$item ) {
        this.ing$amt$qty = ing$amt$qty;
        this.ing$amt$unit = ing$amt$unit;
        this.ing$item = ing$item;
      }
    }
}
```

源代码 8.1 com/tutego/exercise/xml/RecipeMLwriterDemo.java

一个配方由两个类 Recipe 和 Ingredient 表示。它是一个嵌套类型，Ingredient 很好表达了 Recipe 和 Ingredient 两种类型的相互关系。原则上，人们可以给每个子元素定义一个单独的类，但这对于建议的解决方案来说太多了。因此，带有美元的变量名表达了类型的层次性。

generateRandomRecipe( ) 创建一个配方：

```java
static Recipe generateRandomRecipe() {
```

```
  Recipe.Ingredient ingredient1 =
    new Recipe.Ingredient( "30", "cups", "fat" );
  Recipe.Ingredient ingredient2 = new Recipe.Ingredient( "1", "kg", "sugar" );
  return new Recipe( "Fat Jam", Arrays.asList( "Canning", "Preserves" ), "8",
    Arrays.asList( ingredient1, ingredient2 ),
    Arrays.asList( "Start", "End" ) );
}
```

源代码 8.2 com/tutego/exercise/xml/RecipeMLwriterDemo.java

通过参数化构造函数构建两个成分，并创建两个 Ingredient 对象，然后通过参数化的构造函数建立一个配方，两个 Ingredient 对象以数组的形式被传递。

RecipeMLwriter 类和一个公共方法 writeRecipeAsXml(…) 负责编写：

```
class RecipeMLwriter {
  public static void writeRecipeAsXml( OutputStream outputStream,
    Recipe recipe )
      throws XMLStreamException {
    XMLOutputFactory outputFactory = XMLOutputFactory.newFactory();
    XMLStreamWriter writer = outputFactory.createXMLStreamWriter(
      outputStream,
      StandardCharsets.UTF_8.name() );
    writeRecipeElements( writer, recipe );
    writer.close();   // This does not close the underlying output stream
  }

  private static void writeCharacters( XMLStreamWriter writer, String name,
    String string )
      throws XMLStreamException {
    writer.writeStartElement( name );
    writer.writeCharacters( string );
    writer.writeEndElement();
  }

  private static void writeRecipeElements( XMLStreamWriter writer,
    Recipe recipe )
      throws XMLStreamException {
```

```java
    writer.writeStartDocument( "utf-8", "1.0" );
    writer.writeComment( "Recipe" );

    writer.writeStartElement( "recipe" );

    writer.writeStartElement( "head" );

    writeCharacters( writer, "title", recipe.head$title );

    writer.writeStartElement( "categories" );
    for ( String cat : recipe.head$categories )
      writeCharacters( writer, "cat", cat );

    writer.writeEndElement(); // </categories>

    writeCharacters( writer, "yield", recipe.head$yield );

    writer.writeEndElement(); // </head>

    writer.writeStartElement( "ingredients" );
    for ( Recipe.Ingredient ingredient : recipe.ingredients )
      writeIngredientElements( writer, ingredient );
    writer.writeEndElement(); // </ingredients>

    writer.writeStartElement( "directions" );
    for ( String step : recipe.directions )
      writeCharacters( writer, "step", step );
    writer.writeEndElement(); // </directions>

    writer.writeEndElement(); // </recipe>

    writer.writeEndDocument();
  }

  private static void writeIngredientElements( XMLStreamWriter writer,
    Recipe.Ingredient ingredient )
```

```java
    throws XMLStreamException {
  writer.writeStartElement( "ing" );

  writer.writeStartElement( "amt" );
  writeCharacters( writer, "qty", ingredient.ing$amt$qty );
  writeCharacters( writer, "unit", ingredient.ing$amt$unit );
  writer.writeEndElement();   // </amt>

  writeCharacters( writer, "item", ingredient.ing$item );

  writer.writeEndElement();   // </ing>
  }
}
```

源代码 8.3 com/tutego/exercise/xml/RecipeMLwriterDemo.java

从 XMLOutputFactory 的结构开始，XMLStream-Writer 也是通过它来查询的，它使用中心方法 writeStartElement(...)，writeEndElement(...) 和 writeCharacters(...) 来编写 XML 文件。子树是由单个方法来编写的。这比较明晰，因为如果使用一些嵌套，那么每个方法的行数会很多，目前行数也是到了最多数的边界。

可以确认这种编写 XML 文件的方式是高效的，但是如果稍后更改 XML 文件的属性，这种方法也是很不方便的。有一些替代技术可用于自动编写 XML 文件，即 JAXB。

如果你对 RecipeML 的 XML 模式感兴趣，那么可以在 https://github.com/ tranchis/xsd2thrift/blob/master/contrib/recipeml.xsd. 中看到其格式基本已经固定了。

任务 8.1.2：检查所有图片是否有一个 alt 属性

```java
static void reportMissingAltElements( Path path ) {
  try ( InputStream is = Files.newInputStream( path ) ) {
    for ( XMLStreamReader parser =
      XMLInputFactory.newInstance().createXMLStreamReader( is );
      parser.hasNext(); ) {
      parser.next();
      boolean isStartElement = parser.getEventType() ==
        XMLStreamConstants.START_ELEMENT;
      if ( isStartElement ) {
```

```java
        boolean isImgTag = "img".equalsIgnoreCase( parser.getLocalName() );
        if ( isImgTag && ! containsAltAttribute( parser ) )
          System.err.printf( "img does not contain alt attribute:%n%s%n",
            parser.getLocation() );
      }
    }
  }
  catch ( IOException | XMLStreamException e ) {
    throw new RuntimeException( e );
  }
}

private static boolean containsAltAttribute( XMLStreamReader parser ) {
  return IntStream.range( 0, parser.getAttributeCount() )
                  .mapToObj( parser::getAttributeLocalName )
                  .anyMatch( "alt"::equalsIgnoreCase );
}
```

源代码 8.4 com/tutego/exercise/xml/XhtmlHasImgTagWithAltAttribute.java

传递给 createXMLStreamReader(...) 方法的 Path 对象是 InputStream 的基础，我们把它传递给 createXMLStreamReader(...)，以此得到一个带有这个输入流的 XMLStreamReader。不幸的是，迄今为止（从 Java 16 开始）XMLStreamReader 不是 AutoCloseable，因此它不能在 try-with 资源中被关闭。然而，这在读取时并不重要，可以通过一个 try-mit-Ressourcen 块关闭文件的 InputStream。

通过 XMLStreamReader 传递数据看起来总是一样的：has-Next( ) 判断数据流中是否还有标记，如果有，就用 next( ) 获取下一个标记。这类似 Scanner 和 Iterator。对 next( ) 的调用改变了 XMLStreamReader 元素的状态，而 getEventType( ) 返回一个整数，以识别存在的数据。这可以成为文件的开始、处理指令、评论、文本，甚至开始元素。

我们使用常数代替整数，将 XMLStreamReader 接口扩展成 XMLStreamConstants。如果一个元素开始，则它可能是一个 img 元素。因此，getLocalName( ) 向解析器询问元素的名称并将其与 img 进行比较（不区分大小写）。如果结果为真，则表示已经找到了一个 img 标签。现在的问题是，属性 alt 是否也被设置了。这个问题由单独的方法 containAltAttribute(...) 回答。如果 img 标签不带有 alt 属性，就会在标准错误通道上产生一条信息，通过 getLocation( ) 也可以确定确切的位置并在报错信息中指定。

containsAltAttribute(...) 获取 XMLStreamReader 作为参数，并运行从 0 到 getAttributeCount( ) 的所有属性。如果一个属性 alt 存在，则无论如何赋值，该方法都返回 true，否则返回 false。

任务 8.1.3: 编写带有 JAXB 的 Java 对象

```
@XmlRootElement
class Ingredients {
  public Ing[] ing;
}

class Ing {
  public Amt amt;
  public String item;
}

class Amt {
  public int qty;
  public String unit;
}
```

源代码 8.5 com/tutego/exercise/xml/JaxbRecipeML.java

```
Ing ing1 = new Ing();
Amt amt1 = new Amt();
amt1.qty = 3;
amt1.unit = "cups";
ing1.amt = amt1;
ing1.item = "Strawberries";

Ing ing2 = new Ing();
Amt amt2 = new Amt();
amt2.qty = 3;
amt2.unit = "cups";
ing2.amt = amt2;
ing2.item = "Sugar";
```

```
Ingredients ingredients = new Ingredients();
ingredients.ing = new Ing[]{ ing1, ing2 };

JAXB.marshal( ingredients, System.out );
```
源代码 8.6 com/tutego/exercise/xml/JaxbRecipeML.java

使用 JAXB 工作很简单。
1. 编写具有无参数构造函数的类，并使用 setters/getters 或公共对象变量来获取数据。
2. 建立一个对象网格，并用 JAXB.marshal(ingredients, System.out) 把它写到一个输出流中，例如写到控制台中。

出于兼容性的原因，建议解决方案将注释 @XmlRootElement 放在根元素成分上。对于当前的 JAXB 实现来说，这已经没有必要了，但出于兼容性的考虑，这样的解决方案也可以在 Java 8 下工作，因为 Java 8 包含了一个有点老的 JAXB 版本（JAXB RI 2.2.8），目前是 2.4.0。

任务 8.1.4：阅读笑话并开心地笑

在 JAXB 中，JavaBeans 是重点，它使用注解来告诉 JAXB 框架对象如何在 XML 中被映射或 XML 的映射如何在对象上展示。可以手动编写和注释这些 JAXB bean，也可以让它们从模式中生成。任务要求生成这个变体，生成的类 Data 是这样开始的：

```
@XmlAccessorType(XmlAccessType.FIELD)
@XmlType(name = "", propOrder = {
    "category", "type", "flags", "setup", "delivery", "id", "error"
})
@XmlRootElement(name = "data")
public class Data {

    @XmlElement(required = true)
    protected String category;
    @XmlElement(required = true)
```

```
    protected String type;
    ...
}
```

对于用户：

```
try {
  URL url = new URL( "https://sv443.net/jokeapi/v2/joke/Any?format=xml" );
  Data data = JAXB.unmarshal( url, Data.class );
  System.out.println( data.getSetup() );
  System.out.println( data.getDelivery() );
  System.out.printf( "Not Safe for Work?%s%n", data.getFlags().isNsfw() );
  System.out.printf( "Religious?%s%n", data.getFlags().isReligious() );
  System.out.printf( "Political?%s%n", data.getFlags().isPolitical() );
  System.out.printf( "Racist?%s%n", data.getFlags().isRacist() );
  System.out.printf( "Sexist?%s%n", data.getFlags().isSexist() );
}
catch ( MalformedURLException e ) {
  System.err.println( "malformed URL has occurred" );
  e.printStackTrace();
}
catch ( DataBindingException e ) {
  System.err.println( "failure in a JAXB operation" );
  e.printStackTrace();
}
```

源代码 8.7 com/tutego/exercise/xml/JaxbJokeReceiver.java

AXB.unmarshal(...) 允许从来自不同数据源（包括 URL）的 XML 流构造 Java 对象。因此，如果构造一个 URL 对象并将其放在笑话的端点上，unmarshal(...) 将直接返回一个 Data 对象。然后，该数据对象提供不同的 getters 函数，并且可以读出数据。有两种例外情况可能发生：URL 的格式可能是无效的，这将给我们一个 MalformedURLException，或者 XML 格式不能被映射到 JavaBean 上，在这种情况下，结果是带给我们一个 DataBindingException。

任务 8.2.1：黑客新闻：评估 JSON 文件 ★

Jackson 类型的 ObjectMapper 提供 readValue(...)，它将 JSON 文件返回到嵌套键值对的 Map 中。两个建议的解决方案如下：

```java
public static Map<?, ?> news( long id ) {
  try {
    URI uri = URI.create( "https://hacker-news.firebaseio.com/v0/item/" +
      id + ".json" );
    return new ObjectMapper().readValue( uri.toURL(), Map.class );
  }
  catch ( IOException e ) {
    return Collections.emptyMap();
  }
}
```
源代码 8.8 com/tutego/exercise/json/HackerNewsJackson.java

ObjectMapper 对象的 readValue(...) 方法可以在第一个参数中获取 URL，从而直接从网络中获得 JSON 文件并将其转换为 Map。因此，实际上只有一行是必要的。在这种情况下，程序会捕捉到异常并返回一个空 Map。

查看 Jackson 实现的源代码可以发现，它使用 url.openStream() 方法来获取 JSON 数据，但是在 Java 11 中，HTTP-Client-API 更为强大，它涉及认证器、线程池、代理、SSL 上下文等方面的配置。

```java
public static Map<?, ?> news( long id ) {
  HttpClient client = HttpClient.newHttpClient();
  ObjectMapper mapper = new ObjectMapper();

  HttpRequest request = HttpRequest
      .newBuilder( URI.create(
        "https://hacker-news.firebaseio.com/v0/item/" + id + ".json" ) )
      .timeout( Duration.ofSeconds( 5 ) )
      .build();

  try {
    InputStream body = client.send( request,
```

```
          HttpResponse.BodyHandlers.ofInputStream() ).body();
      return mapper.readValue( body, Map.class );
    }
    catch ( IOException | InterruptedException e ) {
      return Collections.emptyMap();
    }
  }
}
```

源代码 8.9 com/tutego/exercise/json/HackerNewsJackson.java

对于 HttpClient，我们访问 HttpClient.newHttpClient( ) 返回到标准配置。该程序还生成 ObjectMapper，无须任何额外的附加功能。HttpRequest 被放在 URL 上，然后再次设置超时。结果应使用 readValue(...) 能够接受的格式。例如 InputStream、字节数组或字符串。InputStream 的优势在于，在最好的情况下，它在读取时需要最少的内存。

任务 8.2.2: 以 JSON 格式读取写入编辑器的配置

JSON 层次结构会自动生成，因为 Settings 引用了一个 Editor 和一个 Workbench：

```
import java.util.*;

public class Settings {

  enum FontWeight {
    normal, bold;
  }

  public static class Editor {
    public String cursorStyle = "line";
    public boolean folding = true;
    public List<String> fontFamily =
      Arrays.asList( "Consolas", 'Courier New', monospace" );
    public int fontSize = 14;
    public FontWeight fontWeight = FontWeight.normal;
  }
```

```java
  public static class Workbench {
    public String colorTheme = "Default Dark+";
    public String iconTheme;
  }

  public Editor editor = new Editor();
  public Workbench workbench = new Workbench();
  public Map<String, String> terminal = new HashMap<>();
}
```

源代码 8.10　com/tutego/exercise/json/Settings.java

Jackson 直接访问对象变量并接收对象的小写标识符。Jackson 直接将列表映射到 JSON 数组上。枚举也是被直接写入的，并可以再次被读出。通过 Map<String, String> 终端，可以使用任何字符串的键值对，这些键值对不被绑定到特殊的对象变量上，而是进入关联内存。另外，为文档的根部做一个特殊的注释（JAXB 中的 @XmlRootElement）是不必要的。Jackson 对类型的要求比 JAXB 少很多。

```java
public class EditorPreferences {

  private static final Path FILENAME = Paths.get(
      /*System.getProperty( "user.home" ),*/ ".editor-configuration.json" );

  private final ObjectMapper jsonMapper = new ObjectMapper();

  private Settings settings = new Settings();

  public EditorPreferences() {
    jsonMapper.enable( SerializationFeature.INDENT_OUTPUT );
    jsonMapper.setSerializationInclusion( JsonInclude.Include.NON_NULL );
  }

  public Settings settings() {
    return settings;
  }

  public Settings load() {
```

```java
    try ( InputStream is = Files.newInputStream( FILENAME ) ) {
      settings = jsonMapper.readValue( is, Settings.class );
      return settings;
    }
    catch ( IOException e ) {
      return settings;
    }
  }

  public void save() {
    try ( OutputStream os = Files.newOutputStream( FILENAME ) ) {
      jsonMapper.writeValue( os, settings );
    }
    catch ( IOException e ) {
      throw new IllegalStateException( e );

    }
  }
}
```
源代码 8.11 com/tutego/exercise/json/EditorPreferences.java

```java
EditorPreferences preferences = new EditorPreferences();
preferences.save();

Settings settings = preferences.load();
settings.editor.fontSize = 22;
settings.terminal.put( "integrated.unicodeVersion", "11" );
preferences.save();
```
源代码 8.12 com/tutego/exercise/json/EditorPreferencesDemo.java

EditorPreferences 有一个构造函数，在构造函数内，通过 ObjectMapper 进行以下配置：

- ▶ 输出要有格式，因为配置文件是为用户制作的，所以配置文件不应该尽可能短以节省所有空格，而是要有中断和插入进行分隔。
- ▶ 如果一个字符串是空的，则 Jackson 写它的方式与引用变量为空的方式相

同。在我们的例子中，不希望是空值；这个属性设置了 SerializationInclusion
（JsonInclude.Include.NON_NULL）。配置提供了通过 ObjectMapper 全局设置
一次或通过注解本地设置的可能性。

在内部，EditorPreferences 创建一个 Settings 对象，该对象从 JSON 文件重构
load( ) 并写入 save( ) 方法。ObjectMapper 对象的 readValue(...) 和 writeValue(...) 负责
实际的映射。

任务 8.3.1: 使用 jsoup 加载维基百科图像

```
String url = "https://de.wikipedia.org/wiki/Wikipedia:Hauptseite";
Document doc = Jsoup.parse( new URL( url ), 1000 /* ms */ );

for ( Element img : doc.select( "img[src~=(?i)\\.(png|gif|jpg)]" ) ) {
  String imgUrl = img.absUrl( "src" );
  String filename = imgUrl.replaceAll( "[^a-zA-Z0-9_.-]", "_" );
  try ( InputStream imgStream = new URL( imgUrl ).openStream() ) {
    Files.copy( imgStream, Paths.get( filename ),
                StandardCopyOption.REPLACE_EXISTING );
  }
}
```

源代码 8.13 com/tutego/exercise/net/WikipediaImageLoader.java

Jsoup 类有静态方法 parse(...)，它可以通过不同的来源建立 HTML 文件。在我
们的例子中，直接选择 URL 对象。当访问网络时，必须给 Jsoup 一个超时时间，
我们将其设置为 1 000 毫秒。parse(...) 方法返回一个 org.jsoup.nodes.Document 对象。
有两种方法可以访问这个文件和提取元素。

- DOM 方法，和 getElementById(String id) 或者 child(int index)。
- Selector-Ausdrücke, wie sie von CSS bekannt sind。

选择器表达方式从 CSS 得到。
建议解决方案和 select(...) 方法一起工作。img[...] 代表所有 img 标签，规范 src~=
通过正则表达式确定什么应该适用于 src 属性，这些字符串匹配在 /(png|gif|jpg)，即文
件的后缀是 .png、.gif 或 .jpg。标志 (?i) 激活搜索，搜索不区分大 / 小写。

select(...) 方法的结果是 Elements 类型，它是 ArrayList 的子类。List 是 Iterable(可迭代的)，它可以方便地使用扩展的 for 循环来运行。列表中的每个元素都是 Element 类型。通过调用 attr("src") 可以询问图片的设定 URL。但是，方法 absUrl(…) 绝对是对于解决 URL 最有用的。

当以后下载图片时，不能直接使用这个 URL 作为文件名，因为那里有非法的符号，会在文件系统中引起问题。字符串方法 replaceAll(...) 返回一个新的整齐的字符串，可以用它作为文件名。下一步是建立一个 URL 对象，为这个图像打开一个输入流，并使用 Files.copy(...) 将其复制到本地文件系统中。我们已经在之前的 ImageDownloader 任务中编写了代码。

任务 8.4.1: 生成带截图的 Word 文件

```
private static final int TOTAL_NUMBER_OF_SCREEN_CAPTURES = 3;
private static final int DURATION_BETWEEN_SCREEN_CAPTURES = 5;
private static final Rectangle SCREEN_SIZE =
  new Rectangle( Toolkit.getDefaultToolkit().getScreenSize() );

private static byte[] getScreenCapture() throws AWTException, IOException {
  BufferedImage screenCapture = new Robot().createScreenCapture( SCREEN_SIZE );
  ByteArrayOutputStream os = new ByteArrayOutputStream();

  ImageIO.write( screenCapture, "jpeg", os );
  return os.toByteArray();
}

private static void appendImage( XWPFDocument doc, byte[] imageBytes )
    throws IOException, InvalidFormatException {
    XWPFRun paragraph = doc.createParagraph().createRun();
    paragraph.addPicture( new ByteArrayInputStream( imageBytes ),
      Document.PICTURE_TYPE_JPEG,
      UUID.randomUUID().toString(),
      Units.toEMU( SCREEN_SIZE.width / 100. * 20 ),
      Units.toEMU( SCREEN_SIZE.height / 100. * 20 ) );
    paragraph.addBreak();
}

public static void main( String[] args ) throws Exception {
```

```java
try ( XWPFDocument xwpfDocument = new XWPFDocument() ) {
  for ( int i = 0; i < TOTAL_NUMBER_OF_SCREEN_CAPTURES; i++ ) {
    appendImage( xwpfDocument, getScreenCapture() );
    TimeUnit.SECONDS.sleep( DURATION_BETWEEN_SCREEN_CAPTURES );
  }

  Path tempFile = Files.createTempFile( "screen-captures", ".docx" );
  try ( OutputStream out = Files.newOutputStream( tempFile ) ) {
    xwpfDocument.write( out );
  }
  System.out.println( "Written to " + tempFile );
}
```

源代码 8.14 com/tutego/exercise/fileformat/ScreenCapturesInDocx.java

该解决方案包含 3 个方法。第一个方法是 getScreenCapture( ) 返回一个 byte[]，其中包含屏幕内容的 JPEG 格式。Java 可以通过 Robot 类来实现这一点，该类的目的是实现自动化。整个屏幕尺寸的 createScreenCapture(...) 的结果是 BufferedImage 类型，它是一种内部图像格式。为了将其转换为 JPEG 格式，程序使用 ImageIO.write(...) 方法，该方法首先将 BufferedImage 写入 ByteArrayOutputStream，然后将其转换为 byte 数组并返回。

第二个方法是 appendImage(...)，它将图像附加到现有的 XWPFDocument。由于每张图片都有自己的段落，所以先建立一个段落，然后通过 addPicture(...) 添加图片。该方法期望有一个图片的输入流，以及一个独特的标识符和大小信息。图片的比例略有不同。

第三个方法是用 main(...) 打开一个新的 XWPFD 文件，然后获取一个屏幕截图，将其附加到文件上，等待 1 秒钟，然后获取一个屏幕截图，直到达到所需的最大数量。该文件当前仅在内存中。Files.createTempFile(...) 在临时目录中创建一个文件并将 Office 文件写入该文件。

任务 8.5.1：播放 ZIP 档案中的昆虫声音

```
Path path = new TPath( filename );

List<Path> wavFiles = new ArrayList<>();
try ( DirectoryStream<Path> entries = Files.newDirectoryStream( path ) ) {
  entries.forEach( wavFiles::add );
}
```

```
while ( true ) {
  Path randomWavFile = wavFiles.get(
    ThreadLocalRandom.current().nextInt( wavFiles.size() ) );
  try ( InputStream fis = Files.newInputStream( randomWavFile );
        BufferedInputStream bis = new BufferedInputStream( fis );
        // for mark/reset support
        AudioInputStream ais = AudioSystem.getAudioInputStream( bis ) ) {
    Clip clip = AudioSystem.getClip();
    clip.open( ais );
    clip.start();
    TimeUnit.MICROSECONDS.sleep( clip.getMicrosecondLength() + 50 );
    clip.close();
  }
}
```

源代码 8.15 com/tutego/exercise/io/TrueZipDemo.java

解决方案由以下部分组成：
1. 从档案 filename 中读取所有音频文件。
2. 选择一个随机的音频文件。
3. 播放音频文件。

TrueZIP 使用它自己的路径实现 TPath 进行工作。构造函数可以传递一个字符串、路径、URI 或文件对象。如果使用 ZIP 档案并建立了 TPath，则 newDirectoryStream(...) 会返回所有的目录内容，将它们缓存在一个 Path 对象的列表中。

无限循环开始并从列表中随机选择一个文件。一个输入流被打开，然后用 BufferedInputStream 进行装饰。这是必要的，因为音频系统在输入流中需要一个特殊的属性，也就是标记被设置。至少在当前版本中，TrueZIP 的输入流不支持这一点。

然后，可以打开 AudioInputStream 并播放该片段。在播放开始后，等待时间（也就是片段的持续时间）可以通过 getMicrosecondLength(...) 查询。

在顶部放置了一个 50 微秒的小缓冲区。片段被关闭，下一次循环执行开始。

第 9 章
用 JDBC 访问数据库

Java 数据库连接（JDBC）为访问不同的关系型数据库提供了一个数据库接口，并允许在关系型数据库管理系统（RDBMS）中执行 SQL 语句。JDBC API 由 JDBC 驱动程序实现。本章提供了一个使用 JDBC API 的示例，允许 CiaoCiao 船长将个人特征和用户信息存储在数据库中以进行海盗约会。

本章使用的数量类型如下：

- java.sql.DriverManager (https://docs.oracle.com/en/java/javase/11/docs/api/java.sql/java/sql/DriverManager.html)
- java.sql.Connection (https://docs.oracle.com/en/java/javase/11/docs/api/ java.sql/java/sql/Connection.html)
- java.sql.SQLException (https://docs.oracle.com/en/java/javase/11/docs/api/ java.sql/java/sql/SQLException.html)
- java.sql.Statement (https://docs.oracle.com/en/java/javase/11/docs/api/java.sql/java/sql/Statement.html)
- java.sql.Statement (https://docs.oracle.com/en/java/javase/11/docs/api/java.sql/java/sql/PreparedStatement.html)
- java.sql.ResultSet (https://docs.oracle.com/en/java/javase/11/docs/api/java.sql/java/sql/ResultSet.html)
- java.sql.Date (https://docs.oracle.com/en/java/javase/11/docs/api/java.sql/java/sql/Date.html)
- java.sql.ResultSetMetaData (https://docs.oracle.com/en/java/javase/11/docs/api/ java.sql/java/sql/ResultSetMetaData.html)

9.1 数据库管理系统

使用 JDBC 进行练习需要一个数据库管理系统、一个数据库和数据。这些任务可以通过任何 RDBMS 来实现，因为所有主要的数据库管理系统都有 JDBC 驱动程序，而且访问看起来总是一样的。本章使用紧凑型数据库管理系统 H2。

> **提示：**
> 有些图形化的工具可以显示表格并简化 SQL 查询的输入。开发环境通常有插件。IntelliJ Ultimate 默认带有一个数据库编辑器（对于免费的社区版，例如：https://plugins.jetbrains.com/plugin/1800-database-navi- gator. Für Eclipse existieren unterschiedliche Plugins, von der Eclipse Foundation selbst das Eclipse Data Tools Platform (DTP) Project unter https://www.eclipse.org/datatools/downloads.php)。

9.1.1 准备 H2 ★

H2 非常紧凑，以至于数据库管理系统、JDBC 驱动程序和一个小的管理界面都打包在一个 JAR 存档中。在 Maven POM 中包含以下依赖项：

```xml
<dependency>
  <groupId>com.h2database</groupId>
  <artifactId>h2</artifactId>
  <version>1.4.200</version>
</dependency>
```

9.2 数据库查询

每个数据库查询都要经过以下步骤：
1. 通过建立连接启动数据库访问；
2. 提交指令；
3. 收集结果。

9.2.1 查询所有注册的 JDBC 驱动程序 ★

Java 6 引入了 Service Provider API，当代码位于类路径下并在一个特殊的文本文件中时，它可以自动运行代码。JDBC 驱动程序使用 Service Provider API，自动登录 DriverManager 类型的控制中心。

任务

通过 DriverManager 查询所有注册的 JDBC 驱动程序，并在屏幕上显示类名。

9.2.2 建立数据库并执行 SQL 脚本 ★

CiaoCiao 船长想要将有关海盗的信息存储在关系型数据库中。初稿设计方案是存储海盗的昵称、电子邮件地址、佩剑的长度、出生日期以及一段简短的描述。对数据库建模后，编写一个构建表的 SQL 脚本：

```
DROP ALL OBJECTS;

CREATE TABLE Pirate (
    id              IDENTITY,
    nickname        VARCHAR(255) UNIQUE NOT NULL,
    email           VARCHAR(255) UNIQUE NOT NULL,
    swordlength     INT,
    birthdate       DATE,
    description     VARCHAR(4096)
);
```

第一条 SQL 语句删除 H2 中的所有条目，然后 CREATE TABLE 创建一个具有不同列和数据类型的新表。每个海盗都有一个由数据库分配的唯一 ID，这被称为自动生成的密钥。

信息：

书中的 SQL 语言遵循以下命名约定：
▶ SQL 关键字一致大写。
▶ 表名是单数，以大写字母开头，就像 Java 中的类名以大写字母开头一样。
▶ 表的行列名小写。

Java SE 程序使用 DriverManager 通过 getConnection(...) 方法建立连接。JDBC URL 包含有关数据库和连接详情（例如服务器和端口）的信息。在 H2 的情况下，

如果不联系任何服务器，则 JDBC URL 很简单，但 RDBMS 将成为自己应用程序的一部分：

```
try ( Connection connection = DriverManager.getConnection(
  "jdbc:h2:./pirates-dating" ) {
  ...
}
```

如果数据库 pirates-dating 不存在，那么它将被创建。getConnection(...) 返回连接。连接必须始终是被关闭。从上面的代码中可以看出，try-with-resources 接收了关闭。

如果整个 RDBMS 作为其自身应用程序的一部分运行，则这被称为嵌入式模式。在嵌入式模式下，启动的 Java 应用程序单独使用该数据库，而其他多个 Java 程序无法连接到该数据库。多个连接只可能在一个数据库服务器上实现。H2 也可以做到这一点。有兴趣的读者可以通过 H2 网站了解如何进行：https://www.h2database.com/html/tutorial.html。

任务：
- 在 Maven 项目的资源目录中放置一个文件 creat-table.sql。将 SQL 脚本复制到文件中。
- 创建一个新的 Java 类并从类路径加载 SQL 脚本。
- 建立一个与数据库的连接，并加载运行 SQL 脚本。

最后，可以通过一个命令行工具查询数据库：

```
$ java -cp h2-1.4.200.jar org.h2.tools.Shell -url jdbc:h2:C:\pfad\zum\ordner\pirates-dating
```

```
Welcome to H2 Shell 1.4.200 (2019-10-14)
Exit with Ctrl+C
Commands are case insensitive; SQL statements end with ';'
help or ?         Display this help
list              Toggle result list / stack trace mode
maxwidth          Set maximum column width (default is 100)
autocommit        Enable or disable autocommit
history           Show the last 20 statements
```

```
quit or exit     Close the connection and exit
sql> SHOW TABLES;
TABLE_NAME  | TABLE_SCHEMA
PIRATE      | PUBLIC
(1 row, 15 ms)
sql> exit
Connection closed
```

> **提示：**
> 访问 Statement 的 execute(...) 方法。

9.2.3 向数据库写入数据 ★

到目前为止建立的数据库不包含任何条目。在以下三个程序中数据记录将被添加。SQL 使用 INSERT 写入数据记录。将一个新的海盗数据通过下面的 SQL 程序写入数据库：

```
INSERT INTO Pirate (nickname, email, swordlength, birthdate, description)
VALUES ('CiaoCiao', 'captain@goldenpirates.faith', 18, DATE '1955-11-07',
 'Great guy')
```

规范中明确缺少主键 ID，因为该列被自动分配了一个唯一值。

任务：

▶ 建立一个与数据库的新连接，创建一个 Statement 对象，并向数据库发送带有 executeUpdate(...) 的 INSERT INTO 语句。

▶ JDBC 驱动程序可以提供生成的密钥。添加第二个海盗数据，并将生成的密钥（即一个 long 数据）在屏幕上显示。方法 executeUpdate(...) 返回一个 int 数据，这对所执行的指令有什么意义？

9.2.4 在批处理模式下向数据库添加数据 ★

如果要执行多个 SQL 语句，可以将它们批量 (Batch) 收集。首先，所有 SQL 语句都被收集起来，然后打包传送到数据库。这样，JDBC 驱动程序不必为每个查询通过网络访问数据库。

任务：

▶ 创建一个新的类并将下列数组放入程序：

```
String[] values = {
    "'anygo', 'amiga_anker@cutthroat.adult', 11, DATE '2000-05-21',
    'Living the dream'", "'SweetSushi', 'muffin@berta.bar', 11,
    DATE '1952-04-03', 'Where are all the bad boys?'", "'Liv Loops',
    'whiletrue@deenagavis.camp', 16, DATE '1965-05-11', 'Great guy'" };
```

▶ 从 SQL-INSERT 数组中的信息生成指令，使用 addBatch(...) 将它们添加到 Statement 中，并使用 executeBatch() 发送指令。

▶ executeBatch() 返回一个 int 数组，其中包括哪些内容？

9.2.5 用准备好的指令添加数据 ★

第三种添加数据的方式在实践中是最高效的。它使用数据库的属性和准备好的指令。Java 通过 PreparedStatement 数据类型来支持它。首先将带有占位符的 SQL 语句发送到数据库，然后将数据单独传输。这样做有两个好处：与数据库通信的数据量少，同时 SQL 指令一般由数据库解析和准备，因此执行速度更高。

任务：

▶ 创建一个新的类并在代码中添加以下声明：

```
List<String[]> data = Arrays.asList(
    new String[]{ "jacky overflow", "bullet@jennyblackbeard.red", "17",
                  "1976-12-17", "If love a crime" },
    new String[]{ "IvyIcon", "array.field@graceobool.cool", "12",
                  "1980-06-12", "U&I" },
    new String[]{ "Lulu De Sea", "arielle@dirtyanne.fail", "13",
                  "1983-11-24", "You can be my prince" }
);
```

▶ 用以下 SQL 语句创建一个准备好的语句字符串：

```
String preparedSql = "INSERT INTO Pirate " +
    "(nickname, email, swordlength, birthdate, description) " +
    "VALUES (?, ?, ?, ?, ?)";
```

▶ 查看数据列表，填写 PreparedStatement，然后发送数据。

▶ 所有添加操作都应在一个大事务块中完成。

9.2.6 数据查询

通过我们的操作，已经把不同的行放进了数据库，现在是把它们读出来的时候了！

任务

- 通过 executeQuery(...) 把 SELECT nickname, swordlength, birthdate FROM Pirate 发送到数据库。
- 读取结果并在屏幕上显示海盗的昵称、佩剑的长度和出生日期。

9.2.7 以交互方式滚动 ResultSet ★

在许多数据库中，Statement 可以按以下方式配置。

- ResultSet 不仅可以被读取，还可以被修改，这样数据就可以很容易地被写回数据库。
- 结果集上的光标不仅可以通过 next( ) 向下移动，还可以任意定位或相对向上设置。

CiaoCiao 船长想在一个互动应用程序中滚动浏览数据库中的所有海盗信息。

任务：

- 在开始时，应用程序应显示记录数据的数量。
- 交互式应用程序侦听控制台输入。d (down) 或 n (next) 应该用下一行填充 ResultSet，u (up) 或 p (previous) 用前一行填充。输入之后，应该输出海盗的昵称，其他细节不要求。
- 注意：next( ) 不能跳到最后一行之后，而 previous( ) 不能跳到第一行之前。

9.2.8 海盗存储库

每个主要应用程序都以某种方式利用外部数据。从领域驱动的设计来看，有一个存储库（Repositorys）的概念。存储库提供所谓的 CRUD 操作：创建、读取、更新、删除。存储库是业务逻辑和数据存储之间的中介。Java 程序只能使用对象，存储库将 Java 对象映射到数据存储，反之，将关系型数据库中的本地数据转换为 Java 对象。在最好的情况下，业务逻辑根本不知道 Java 对象是以何种格式存储的。为了在业务逻辑和数据库之间交换对象，我们要使用 Java 的海盗类。被映射到关系型数据库的对象在 Java 的行话中被称为 Entity-Bean。

```
class Pirate {
  public final Long id;
  public final String nickname;
  public final String email;
  public final int swordLength;
```

```java
  public final LocalDate birthdate;
  public final String description;

  public Pirate( Long id, String nickname, String email, int swordLength,
                 LocalDate birthdate, String description ) {
    this.id = id;
    this.nickname = nickname;
    this.email = email;
    this.swordLength = swordLength;
    this.birthdate = birthdate;
    this.description = description;
  }

  @Override public String toString() {
    return "Pirate[" + "id=" + id + ", nickname='" + nickname + '\"'
        + ", email='" + email + '\"' + ", swordLength=" + swordLength
        + ", birthdate=" + birthdate + ", description='" + description
        + '\"' + ']';
  }
}
```

源代码 9.1 com/tutego/exercise/jdbc/PirateRepositoryDemo.java

业务逻辑通过存储库获取或写入数据。这些操作中的每一个都由一个方法来表达。每个存储库看起来都有点不同，因为业务逻辑想从数据存储中查询和写入不同信息。

任务：
应用程序的建模表明，一个海盗存储库 PirateRepository 是必要的，并且必须提供三种方法。

- List<Pirate> findAll( )：提供数据库中所有海盗的名单。
- Optional<Pirate> findById(long id)：返回一个基于 ID 的海盗，如果数据库中没有该 ID 的海盗，则返回 Optional.empty( )。
- Pirate save(Pirate pirate)：保存或更新海盗。如果海盗还没有主键，即如果 id == null，则应该通过 SQL-INSERT 将海盗写入数据库。如果海盗有一个主键，那么海盗已经在数据库中被保存一次，必须使用 save(…) 方法来代替 SQL-UPDATE 更新。save(…) 方法以始终对具有 set 键的 Pirate 对象进行响应。

开发 PirateRepository 之后，应该可以进行以下操作：

```
PirateRepository pirates = new PirateRepository( "jdbc:h2:./pirates-dating" );
pirates.findAll().forEach( System.out::println );
System.out.println( pirates.findById( 1L ) );
System.out.println( pirates.findById( -1111L ) );
Pirate newPirate = new Pirate(
    null, "BachelorsDelight", "GoldenFleece@RoyalFortune.firm", 15,
    LocalDate.of( 1972, 8, 13 ), "Best Sea Clit" );
Pirate savedPirate = pirates.save( newPirate );
System.out.println( savedPirate );
Pirate updatedPirate = new Pirate(
    savedPirate.id, savedPirate.nickname, savedPirate.email,
    savedPirate.swordLength + 1, savedPirate.birthdate,
    savedPirate.description );
pirates.save( updatedPirate );
pirates.findAll().forEach( System.out::println );
```
源代码 9.2 com/tutego/exercise/jdbc/PirateRepositoryDemo.java

9.2.9 查询列元数据

通常在 Java 程序中，数据库的模式是已知的，所有列都可以在查询中被单独评估。然而，在有些查询和建模时，事先并不知道列的数量。在查询完成后，JDBC 可以请求一个 ResultSetMetaData，让它提供关于列的总数和各列的数据类型的信息。

任务：

- 编写一个方法 List<Map<String, Object>> findAllPirates( )，并返回一个关联存储的列表。列表中的小 Map 对象通过将列名连接到列中条目的关联内存来存储行内容。
- 执行 SQL 查询：SELECT * FROM Pirat。

9.3 建议解决方案

任务 9.2.1: 查询所有注册的 JDBC 驱动程序

```
Collections.list( DriverManager.getDrivers() ).forEach(
```

```
driver -> System.out.println( driver.getClass().getName() )
);
```

源代码 9.3 com/tutego/exercise/jdbc/DriverInfo.java

静态方法 DriverManager.getDrivers( ) 返回一个 Enumeration <Driver>。由于我们不想手动运行 Enumeration，所以通过 Collections.list(...) 将这个元素放在一个列表中。列表是可迭代的（Iterable），因此可以用一个 Consumer 在所有已安装的驱动程序上紧凑地运行，并输出类的名称。

输出的结果如下：

```
org.h2.Driver
```

> **重点：**
> 如果此时输出是空，则这是一个错误，同时下面的程序将无法运行，因为缺少 H2 驱动。在这种情况下需要再次检查依赖性。

任务 9.2.2：建立数据库并执行 SQL 脚本

```
String sql = "";
try ( Scanner scanner =
      new Scanner( CreateTable.class.getResourceAsStream(
      "create-table.sql" ), "UTF-8" ) ) {
  sql = scanner.useDelimiter( "\\z" ).next();
}

try ( Connection connection = DriverManager.getConnection(
      "jdbc:h2:./pirates-dating" );
      Statement statement = connection.createStatement() ) {
  statement.execute( sql );
}
catch ( SQLException e ) { e.printStackTrace(); }
```

源代码 9.4 com/tutego/exercise/jdbc/CreateTable.java

为了让 Java 将 SQL 文件作为资源从类路径中加载，它要进入 Maven 下的 src/main/resources 文件夹。与其把 SQL 文件放在根目录中，不如把它放在与类相同的目录中。在建议解决方案中，该类完全被限定为 com.tutego.exercise.jdbc.
CreateTable，这意味着 create-table.sql 在对称目录 src/main/resources/com/tutego/
exercise/jdbc/create-table.sql 中。

使用 XXX.class.getResource-AsStream(...) 从类路径访问资源，在我们的例子中，
不必关注类加载器。如果 Java 程序以其他方式从类路径加载资源，则这可能是相
关的。如果找不到资源，则 getResourceAsStream(…) 返回一个 InputStream 或 null-
程序保存查询；如果文件不存在，则返回一个 NullPointerException 也很好。

扫描器 (Scanner) 帮助输入文件。在构造函数中，InputStream 被传递并命名了
编码。

在输入时，分隔符被设置为输入的末尾，这样一个简单的调用 next( ) 就可以
输入整个文件。

这是一个很好的技巧，从 Java 9 开始，new String(inputStream. readAllBytes( ),
StandardCharsets.UTF_8) 也可以读取所有字节并建立一个字符串；不推荐使用
StringBufferInputStream。我们不需要一个 catch 块，因为 Scanner 在 close( ) 过程中
不会抛出 IOException 异常。情况会有所不同，如果用 getResourceAsStream(...) 方法
初始化的 InputStream 变量的声明在 try 块中，那么就必须有一个 catch 块，因为通
常 InputStream 使用 close( ) 方法触发抛出 IOException 异常，加载 SQL 脚本后可以
执行。建立数据库连接，并请求一个声明。声明对象提供了 execute(String) 方法，
它可以执行任何 SQL。该方法很少使用，因为通常返回值是相关的，例如改变行
数，或者在 SELECT 的情况下返回结果。

JDBC API 通过 SQLException 报告错误，这是一个检查过的异常状况。在建议
解决方案中捕获该异常，并在命令行中报告错误。JDBC 中的异常处理要复杂一
些，因为 SQLException 对象可以在一个链中包含其他的异常。此外，SQLException
对象包含一个记录确切错误的状态代码。以下解决方案在某种程度上进行了简化，
因为它们没有用于处理 SQLException 的 catch 块，而是用 throws 将异常传递给方法
调用者。

任务 9.2.3：向数据库写入数据

```
try ( Connection connection = DriverManager.getConnection(
    "jdbc:h2:./pirates-dating" );
    Statement statement = connection.createStatement() ) {

  String sql1 =
    "INSERT INTO Pirate " +
    "(nickname, email, swordlength, birthdate, description) " +
```

```
        "VALUES ('CiaoCiao', 'captain@goldenpirate.faith', 18, 
            DATE '1955-11-07', 'Great guy')";
    statement.executeUpdate( sql1 );

    String sql2 = 
        "INSERT INTO Pirate " +
        "(nickname, email, swordlength, birthdate, description) " +
        "VALUES ('lolalilith', 'fixme@bumblebee.space', 12, DATE '1973-07-20', " +
        "I'm 99% perfect')";
    int rowCount = statement.executeUpdate( sql2, Statement.RETURN_GENERATED_KEYS );
    if ( rowCount != 1 )
        throw new IllegalStateException( "INSERT didn't return a row count of 1" );

    ResultSet generatedKeys = statement.getGeneratedKeys();
    if ( generatedKeys.next() )
        System.out.println( generatedKeys.getLong( 1 ) );
}
```
源代码 9.5 com/tutego/exercise/jdbc/InsertData.java

每当执行 INSERT、UPDATE 或 DELETE 等 SQL 语句时，executeUpdate（...）都会被使用。executeUpdate（...）的第一个变量获取 SQL 字符串并执行它。在第二次调用 executeUpdate（...）时，Statement.RETURN_GENERATED_KEYS 作为第一个参数被传递，数据库会指示传输生成的密钥。

在第二个 executeUpdate(...) 中，程序记录了返回值，它说明了修改数据记录的数量。如果 SQL 语句没有返回，则返回值为 0。在示例中，如果 INSERT 插入一条新记录，则返回值为 1，可以将 1 确认为有新的记录输入。

返回的不是生成的密钥。例如，键是字符串，而 executeUpdate(...) 的返回类型总是 int 也是不可能的。通过 Statement 对象在第二个语句 getGenerated-Keys() 中查询生成的密钥。该方法返回一个 ResultSet，稍后也会看到 SELECT 语句的情况。next() 方法确定是否有另一行，并将信息填入 ResultSet，第一列包含生成的 long 类型的主键。如果由于 next() 返回 false 而无法区分大小写，则没有键可以请求。

任务 9.2.4：在批处理模式下向数据库添加数据

```
try ( Connection connection = DriverManager.getConnection(
```

第 9 章　用 JDBC 访问数据库 | 301

```java
            "jdbc:h2:./pirates-dating" ) ) {
  connection.setAutoCommit( false );

  String sqlTemplate = "INSERT INTO Pirate " +
    "(nickname, email, swordlength, birthdate, description) " +
    "VALUES (%s)";

  String[] values = {
    "'anygo', 'amiga_anker@cutthroat.adult', 11, DATE '2000-05-21', " +
    "'Living the dream'",
    "'SweetSushi', 'muffin@berta.bar', 11, DATE '1952-04-03', " +
    "'Where are all the bad boys?'",
    "'Liv Loops', 'whiletrue@deenagavis.camp', 16, DATE '1965-05-11', " +
    "'Great guy'" };

  try ( Statement statement = connection.createStatement() ) {
    for ( String value : values )
      statement.addBatch( String.format( sqlTemplate, value ) );

    int[] updateCounts = statement.executeBatch();
    connection.commit();
    System.out.println( Arrays.toString( updateCounts ) );
  }
}
```

源代码 9.6 com/tutego/exercise/jdbc/BatchInsert.java

变量 sqlTemplate 包含一个格式化字符串，随后任何 VALUES 都可以加入这个字符串。通过一个扩展的 for 循环，程序在数组的条目上运行，并将 SQL 模板与数据连接起来。得到的字符串将被传递给 addBatch(...)。executeBatch(...) 方法运行所收集的 SQL 语句，其返回值是一个数组，该数组所包含的元素数量正好是该批次中执行的 SQL 语句的数量。

在数组中，单元格包含了修改的行数，就像之前看到的 executeUpdate(...) 方法一样。

建议解决方案还进一步重置了 Auto-com-mit Mode。在默认情况下，JDBC 驱动程序将它发送的每个 SQL 语句都放在它自己的事务块中。在批处理中，不确定是进行批处理还是仅处理其中的某部分，或者每个单独的 SQL 语句在一个事务中进行。Java 文档中写道: "The commit behavior of executeBatch is always implementation-defined when an error occurs and auto-commit is true."

在建议的解决方案中，批处理应在事务中进行。为此，必须首先关闭自动提交模式。通过 executeBatch( ) 发送后，事务通过 commit( ) 完成。如果没有异常发生，则所有语句在通过 commit( ) 触发之后都会被提交。

任务 9.2.5: 用准备好的指令添加数据

```
String preparedSql = "INSERT INTO Pirate " +
    "(nickname, email, swordlength, birthdate, description) " +
    "VALUES (?, ?, ?, ?, ?)";

try ( Connection connection = DriverManager.getConnection(
        "jdbc:h2:./pirates-dating" );
      PreparedStatement preparedStatement = connection.prepareStatement(
        preparedSql ) ) {

  connection.setAutoCommit( false );

  List<String[]> data = Arrays.asList(
      new String[]{ "jacky overflow", "bullet@jennyblackbeard.red", "17",
        "1976-12-17", "If love a crime" },
      new String[]{ "IvyIcon", "array.field@graceobool.cool", "12",
        "1980-06-12", "U&I" },
      new String[]{ "Lulu De Sea", "arielle@dirtyanne.fail", "13",
        "1983-11-24", "You can be my prince" }
  );

  for ( String[] elements : data ) {
    preparedStatement.setString( /* nickname */ 1, elements[ 0 ] );
    preparedStatement.setString( /* email */ 2, elements[ 1 ] );
    preparedStatement.setInt( /* swordlength */ 3, Integer.parseInt(
      elements[ 2 ] ) );
    preparedStatement.setDate( /* birthdate */ 4, Date.valueOf(
      elements[ 3 ] ) );
    preparedStatement.setObject( /* description */ 5, elements[ 4 ] );
```

```
      preparedStatement.executeUpdate();
  }

  connection.commit();
}
```

源代码 9.7 com/tutego/exercise/jdbc/PreparedInsert.java

使用 PreparedStatement 的流程总是一样的。首先，建立一个 PreparedStatement，下面总是用到它。不同的 setXXX(...) 方法占据了准备好的 SQL 语句中的占位符。每个 "?" 都有一个以 1 开头的索引来标识。例如，setString(1, ...) 将海盗的昵称分配给第一个问号。顺序完全是问号的顺序，而不是列的顺序。在我们的 SQL 语句中，没有出现列的 ID。

在填充每个 "?" 参数后，准备好的语句通过 execute-Update( ) 方法发送。也可以查询自动生成的密钥。我们将在以后的任务中再讨论这个问题。

所有操作都要在一个事务性的块中再次被执行。然后，自动提交模式再次被关闭，在事务运行结束时，事务通过 commit( ) 确认。

任务 9.2.6: 数据查询

```
String sql = "SELECT nickname, swordlength, birthdate FROM Pirate";
try ( Connection connection = DriverManager.getConnection(
   "jdbc:h2:./pirates-dating" );
      Statement statement = connection.createStatement();
      ResultSet resultSet = statement.executeQuery( sql ) ) {
  while ( resultSet.next() ) {
    String nickname = resultSet.getString( /* nickname column */1 );
    int swordlength = resultSet.getInt( "swordlength" );
    Date birthdate = resultSet.getDate( "birthdate" );
    System.out.printf( "%-20s%-20s%10d%n",
      nickname,
      birthdate.toLocalDate().format( DateTimeFormatter.ofLocalizedDate(
         FormatStyle.LONG ) ),
      swordlength );
  }
}
```

源代码 9.8 com/tutego/exercise/jdbc/Select.java

Statement 对象的 executeXXX( ) 方法返回不同的结果类型。在 executeQuery(...) 的情况下，结果是一个 ResultSet。该对象允许对行进行访问。一个 ResultSet 只包含一行的记录信息。

方法调用 next( ) 把一种光标放置在结果集的下一行，并返回一个布尔值，以说明下一行是否可以被读取。如果 next( ) 返回 true，那么 ResultSet 包含了一行的信息。

可以使用两种不同的方法读取列的分配：使用列索引——在 SQL 中总是从 1 开始，或者使用列名。除此以外，还有不同的 getXXX(...) 方法来转换类型。对于所有 SQL 数据类型，都有相应的 Java 方法给我们提供相应的 Java 类型。例如，getString(...) 方法为文本列返回一个字符串对象。JDBC 驱动程序执行各种转换，例如 getString(...) 适用于任何 SQL 列类型。在 java.sql 包中，Java 有 3 种自己的数据类型用于 SQL 日期和时间值：Date，Time（时间）和 Timestamp（日期和时间）。这些数据类型允许转换为 Java Date Time API 的数据类型。它向 getDate(...) 返回一个 java.sql.Date 对象，该对象将 toLocalDate( ) 转换为已知的 LocalDate，可以使用常用 API 对其进行格式化。

任务 9.2.7：以交互方式滚动 ResultSet

```
int NICKNAME_COLUMN = 2;
String sql = "SELECT * FROM Pirate ORDER BY nickname";
try ( Connection connection = DriverManager.getConnection(
    "jdbc:h2:./pirates-dating" );
     Statement statement = connection.createStatement(
       ResultSet.TYPE_SCROLL_SENSITIVE, ResultSet.CONCUR_READ_ONLY );
     ResultSet srs = statement.executeQuery( sql ) ) {

  if ( srs.last() )
    System.out.printf( "%d rows%n", srs.getRow() );

  srs.absolute( 1 );
  System.out.println( srs.getString( NICKNAME_COLUMN ) );

  for ( String input;
      !(input = new Scanner( System.in ).next()).equals( "q" ); ) {
    switch ( input.toLowerCase() ) {
      case "u": case "p":
```

```java
        if ( srs.isFirst() ) System.out.println( "Already first" );
        else srs.previous();
        break;
      case "d": case "n":
        if ( srs.isLast() ) System.out.println( "Already last" );
        else srs.next();
        break;
    }

    System.out.println( srs.getString( NICKNAME_COLUMN ) );
  }
}
```

源代码 9.9 com/tutego/exercise/jdbc/ScrollableResultSet.java

在移动 ResultSet 之前,必须正确初始化 Statement 对象:

```
createStatement( ResultSet.TYPE_SCROLL_SENSITIVE,
  ResultSet.CONCUR_READ_ONLY );
```

Javadoc 将参数命名为 resultSetType 和 resultSetConcurrency。常量来自 ResultSet,TYPE_SCROLL_SENSITIVE 配置可以在 Resultset 中自由移动光标。并非每个数据库和每个数据库驱动程序都支持此属性,因此可能会发生 SQLFeatureNotSupportedException。

如果成功建立了 Statement 对象,就可以用 SELECT 语句来建立一个 ResultSet。

第一个关于元素总数的问题也可以通过移动光标来实现。如果调用 ResultSet 对象的 last( ),则它会将光标设置到最后一个元素。然后,getRow( ) 方法返回当前行,在我们的例子中是结果集中的记录数。调用 absolute(1) 将光标调回第一行。ResultSet 总是用当前数据填充。如果访问海盗的昵称的列,则结果正是光标当前所处的行的内容。

对于交互式使用,Scanner 有助于控制台的输入。只要用户不按 Q 键(退出),循环就会一直执行。在用户按 U 或 P 键的情况下,他是希望将光标向上移动。当然有可能这时光标不在第一行。这就需要使用 isFirst( ) 检查。当用户按 D 或 N 键时,也会进行对比测试。可以用 next( ) 将光标移动到下一行,除非光标已经在最后一行。

任务 9.2.8：海盗存储库

在开始实际执行之前，应该考虑可能出现的例外情况，即使每个 SQL 查询都应该成功。使所有 JDBCAPI 查询都返回检查过的异常是不现实的。此外，存储库应该隐藏存储技术，这就是 SQLException 不能很好处理的地方。所选择的解决方案引入了一个新类 DataAccessException：

```
class DataAccessException extends RuntimeException {
  public DataAccessException( Throwable cause ) { super( cause ); }
  public DataAccessException( String message ) { super( message ); }
}
```

源代码 9.10 com/tutego/exercise/jdbc/PirateRepositoryDemo.java

DataAccessException 是一个未经检查的异常，它可能包含任意一种异常。
在内部，存储库方法将捕获一个 SQLException 并将其转换为 DataAccessException。
PirateRepository 的范围更广泛，因此首先处理类型声明、常量和构造函数，然后在接下来的步骤中处理各方法。

```
class PirateRepository {

  private static final String SQL_SELECT_ALL =
    "SELECT id, nickname, email, swordlength, birthdate, description FROM
     Pirate";
  private static final String SQL_SELECT_BY_ID =
    "SELECT id, nickname, email, swordlength, birthdate, description FROM
     Pirate WHERE id=?";
  private static final String SQL_INSERT =
    "INSERT INTO Pirate (nickname, email, swordlength, birthdate,
    description) " +
    "VALUES (?, ?, ?, ?, ?)";
  private static final String SQL_UPDATE =
    "UPDATE Pirate SET nickname=?, email=?, swordlength=?,
     birthdate=?, description=? " +
    "WHERE id=?";

  private final String jdbcUrl;
```

```java
  public PirateRepository( String jdbcUrl ) {
    this.jdbcUrl = jdbcUrl;
  }
  // ...
}
```
源代码 9.11 com/tutego/exercise/jdbc/PirateRepositoryDemo.java

PirateRepository 类为 SQL 语句声明了不同的常量。我们需要查询所有海盗、查询给定 ID 的特定海盗、添加新海盗和更新现有海盗都有对应的一条 SQL 语句。只有查询所有海盗的第一条 SQL 语句不使用占位符，否则稍后 SQL 字符串会在 PreparedStatement 中被使用。

程序没有为列声明任何常量，每个索引都被记录在代码中。

构造函数获取 JDBC URL 并将其存储在一个属性中，以便以后单个方法都可以创建到该数据源的连接。如果标识符无效，则后面会给出异常。null 可以提前测试。

我们已经在前面的任务中实现了单个方法的核心，因此这对于 JDBC -CPI 并不新鲜：

Zur ersten Methode findAll( ):
第一个方法 findAll( ):
```java
public List<Pirate> findAll() {
  try ( Connection connection = DriverManager.getConnection( jdbcUrl );
        Statement statement = connection.createStatement();
        ResultSet resultSet = statement.executeQuery( SQL_SELECT_ALL ) ) {
    List<Pirate> result = new ArrayList<>();
    while ( resultSet.next() )
      result.add( mapRow( resultSet ) );
    return result;
  }
  catch ( SQLException e ) {
    throw new DataAccessException( e );
  }
}
```
源代码 9.12 com/tutego/exercise/jdbc/PirateRepositoryDemo.java

Statement 构建完成后，所有数据记录都被选中。while 循环在结果集上运行并调用它自己的私有方法 map Row(...)，它将 ResultSet 中的一行转移到一个 Pirate 对象上。其结果被放在列表中，当光标到达结果的末端时，列表被返回。

```java
private static Pirate mapRow( ResultSet resultSet ) throws  SQLException {
  return new Pirate( resultSet.getLong( /* id */ 1 ),
                     resultSet.getString( /* nickname */ 2 ),
                     resultSet.getString( /* email */ 3 ),
                     resultSet.getInt( /* swordLength */ 4 ),
                     resultSet.getDate( /* birthdate */ 5 ).toLocalDate(),
                     resultSet.getString( /* description */ 6 ) );
}
```

源代码 9.13 com/tutego/exercise/jdbc/PirateRepositoryDemo.java

mapRow(...) 使用已知方法 getLong(...)，getString(...)，getInt(...)，getDate(...) 读取列，并将分配传递给 Pirate 构造函数。Pirate 与 SQL 数据类型 Date 没有关系，因此 toLocalDate() 被转换成一个 LocalDate 对象。替换 ResultSet 对象不是 mapRow 的任务。

findById(...) 方法也可以访问 mapRow(...)：

```java
public Optional<Pirate> findById( long id ) {
  try ( Connection connection = DriverManager.getConnection( jdbcUrl  );
        PreparedStatement prepStatement = connection.prepareStatement(
        SQL_SELECT_BY_ID) ) {
    prepStatement.setLong( 1, id );
    ResultSet resultSet = prepStatement.executeQuery();
    return  resultSet.next() ? Optional.of( mapRow( resultSet ) )
      : Optional.empty();
  }
  catch ( SQLException e ) {
    throw new DataAccessException( e );
  }
}
```

源代码 9.14 com/tutego/exercise/jdbc/PirateRepositoryDemo.java

构建 PreparedStatement 对象后，将发送 SQL 查询并评估 ResultSet。这里有两种可能：ResultSet 可能只有一个结果，也可能没有结果。如果没有结果，则说明该 ID 未被海盗占用，其返回值为 Optional.empty( )。如果 next( ) 返回 true，则 mapRow(...) 从 ResultSet 构造一个 Pirate 对象，该对象进入 Optional 并返回。

save(...) 方法必须使用一个集合 ID 来识别记录是否需要更新或重写。

```java
public Pirate save( Pirate pirate ) {
  try ( Connection connection = DriverManager.getConnection( jdbcUrl ) ) {
    return pirate.id == null ? saveInsert( connection, pirate )
                              : saveUpdate( connection, pirate );
  }
  catch ( SQLException e ) {
    throw new DataAccessException( e );
  }
}

private Pirate saveInsert( Connection connection, Pirate pirate ) throws
  SQLException {
  try ( PreparedStatement prepStatement = connection.prepareStatement(
      SQL_INSERT, Statement.RETURN_GENERATED_KEYS ) ) {
    prepStatement.setString( 1, pirate.nickname );
    prepStatement.setString( 2, pirate.email );
    prepStatement.setInt( 3, pirate.swordLength );
    prepStatement.setDate( 4, Date.valueOf( pirate.birthdate ) );
    prepStatement.setObject( 5, pirate.description );
    prepStatement.executeUpdate();

    ResultSet keys = prepStatement.getGeneratedKeys();
    if ( keys.next() )
      return new Pirate( keys.getLong( 1 ), pirate.nickname, pirate.email,
        pirate.swordLength, pirate.birthdate, pirate.description );
    throw new DataAccessException(
      "Could not retrieve auto-generated key for " + pirate );
  }
}

private Pirate saveUpdate( Connection connection, Pirate pirate ) throws
  SQLException {
  try ( PreparedStatement prepStatement = connection.prepareStatement(
      SQL_UPDATE ) ) {
```

```
        prepStatement.setString( 1, pirate.nickname );
        prepStatement.setString( 2, pirate.email );
        prepStatement.setInt( 3, pirate.swordLength );
        prepStatement.setDate( 4, Date.valueOf( pirate.birthdate ) );
        prepStatement.setObject( 5, pirate.description );
        prepStatement.setLong( 6, pirate.id ); // UPDATE Pirate SET ... WHERE id=?
        prepStatement.executeUpdate();
        return pirate;
    }
}
```

源代码 9.15 com/tutego/exercise/jdbc/PirateRepositoryDemo.java

条件运算符中的大小写区分正是检查这种情况。如果 id == null，则内部 saveInsert(...) 方法负责插入新记录，否则 saveUpdate(...) 更新行。这两个方法都返回一个 Pirate 对象，这就是 save() 返回的内容。

saveInsert(...) 和 saveUpdate(...) 都与 PreparedStatement 一起运行，但是 saveInsert(...) 需要通过请求生成密钥来做更多的工作。由于 Pirate 对象是不可变的，所以通过设置 ID 形成了一个新的 Pirate.saveUpdate(...) 返回传递的 Pirate 对象。

任务 9.2.9: 查询列元数据

```
public static List<Map<String, Object>> findAllPirates() throws SQLException {
    try ( Connection connection = DriverManager.getConnection(
        "jdbc:h2:./pirates-dating" );
        Statement statement = connection.createStatement();
        ResultSet resultSet = statement.executeQuery(
          "SELECT * FROM Pirate" ) ) {
    ResultSetMetaData metaData = resultSet.getMetaData();

    List<Map<String, Object>> result = new ArrayList<>();
    while ( resultSet.next() ) {
      LinkedHashMap<String, Object> map = new LinkedHashMap<>();
      for ( int col = 1, columns = metaData.getColumnCount();
          col <= columns; col++ ) {
        String columnName = metaData.getColumnName( col );
```

```
            map.put( columnName, resultSet.getObject( col ) );
        }
        result.add( map );
    }
    return result;
  }
}
```
源代码 9.16 com/tutego/exercise/jdbc/QueryForListOfMaps.java

查询后，ResultSet 方法 getMetaData( ) 将元数据作为 ResultSetMetaData 对象返回，getColumnCount( ) 返回结果集中的列数。为了将结果集中的数据存储在列表中，我们建立了一个 ArrayList 类型容器。像往常一样对 ResultSet 进行迭代，并将 LinkedHashMap 作为一个映射来构建，以便保留列的顺序。for 循环检查从 1 到 getColumnCount( ) 的列，使用 getObject(...) 方法读取列分配，LinkedHashMap 与列名一起创建一个新条目。在对所有列进行循环后，一行的关联内存包含了所有列分配，映射被放在列表中。对所有行都是重复执行上述流程，最后 findAllPirates( ) 返回一个关联存储的列表。

第 10 章
操作系统的接口

Java 开发人员通常没有意识到有多少 Java 库在操作系统中被使用。举个例子：Java 自动为路径规范设置正确的路径分隔符（/ 或 \），并自动为换行符设置操作系统的常用行尾字符。格式化控制台输入和解析控制台输入时，Java 自动使用设置的操作系统语言。

Java 库不仅在内部使用这些属性，而且每个人都可以使用它们。这些属性隐藏在不同的地方，例如：

- 系统属性。
- 平台 MXBeans，例如具有 getSystemCpuLoad( ) 的 OperatingSystemMXBean。
- java.net.NetworkInterface 中的网卡和 MAC 地址。
- 屏幕分辨率工具包。
- 在 GraphicsEnvironment 中安装的字体。

如果缺少特定信息，则 Java 程序可以调用外部本地程序并与之通信，例如询问更多详细信息。

本章主要关注系统属性和调用外部程序的方法。

本章使用的数据类型如下：

- java.lang.System (https://docs.oracle.com/en/java/javase/11/docs/api/java.base/java/lang/System.html)
- java.util.Properties (https://docs.oracle.com/en/java/javase/11/docs/api/java.base/java/util/Properties.html)
- InputStream (https://docs.oracle.com/en/java/javase/11/docs/api/java.base/java/io/InputStream.html)

- Process (https://docs.oracle.com/en/java/javase/11/docs/api/java.base/java/lang/Process.html)
- ProcessBuilder (https://docs.oracle.com/en/java/javase/11/docs/api/java.base/java/lang/ProcessBuilder.html)

10.1 控制台

Java 程序不仅是算法，它还可以与我们以及操作系统进行互动。前面的两条语句是 System.out.println( ) 和 new Scanner(System.in).next( ) —— 一个屏幕输出和另一个控制台输入，它们是程序和用户之间的经典接口。

10.1.1 彩色的控制台输出 ★

在最开始的 Java 程序中，我们会在控制台输出文本。然而，System.out 和 System.err 上的文本输出显示是黑色的，然后是红色的，还有不同的颜色。

在这个任务中，我们将研究如何解决彩色输出的问题。这背后的解决方案不仅涉及输出文本。

有一些特殊的命令可以改变控制台的颜色，如移动光标、清除屏幕等。ANSI-Escape-Sequenzen 负责这些。它通过 Control Sequence Introducer 开始。

在 Java 中，转义序列可以很容易地用 System.out 或 System.err 编写，并且有 https://github.com/fusesource/jansi 之类的库使用常量和方法来简化这一过程。唯一的问题是不同的控制台不一定能理解所有转义序列，例如 Windows 的 cmd.exe 就无法理解它。然而，对某些颜色的支持在其他控制台中是很普遍的：

- 黑色：\u001B[30m
- 红色：\u001B[31m
- 绿色：\u001B[32m
- 黄色：\u001B[33m
- 蓝色：\u001B[34m
- 洋红色：\u001B[35m
- 青色：\u001B[36m
- 白色：\u001B[37m
- 重置：\u001B[0m

合适的修饰：

- fett: \u001B[1m

- 粗体 : \u001B[1m
- unterstrichen: \u001B[4m
- 下划线 : \u001B[4m
- invertiert: \u001B[7m
- 倒置 : \u001B[7m

任务：

- 创建一个新类 AnsiColorHexDumper 并将以下常量放入该类：
  ```
  public static final String ANSI_RED    = "\u001B[31m";
  public static final String ANSI_GREEN  = "\u001B[32m";
  public static final String ANSI_BLUE   = "\u001B[34m";
  public static final String ANSI_PURPLE = "\u001B[35m";
  public static final String ANSI_CYAN   = "\u001B[36m";
  public static final String ANSI_RESET  = "\u001B[0m";
  ```
- 编写一个新的方法 printColorfulHexDump(Path) 来读取传来的文件并以 hexdump 的形式输出到控制台。十六进制转储是一个文件的十六进制符号的输出，即将 50 4B 03 04 14 00 06 00 08 00 这样的序列写在列中。
- 扩展程序，使颜色通过指示文件中的特定字节展示。例如，ASCII 字母可以以一种颜色出现，数字可以被显示为不同的颜色。

10.2 特性

术语"特性"在 Java 中有多种含义。它代表了 JavaBean 的特性和可用于配置的键值对。当我们在本章中谈论"特性"时，总是指键值对，特别是特性对象所管理的键值对。

10.2.1 Windows 或者 UNIX 或者 macOS？

系统特性包含一系列信息，其中一些特性也可以通过方法访问。Javadoc 可在 https://docs.oracle.com/en/java/javase/11/docs/api/java.base/java/lang/System.html#get-Properties( ) 中列出变量。

任务：

- 通过常量 WINDOWS，MACOS，UNIX，UNKNOWN 创建一个新的枚举类型 OS。
- 增加一个静态方法 current( )，用 System.getProperty("os.name") 读取操作系统的名称，并返回相应的枚举元素作为结果。

10.2.2 统一命令行特性和文件中的特性

特性可以在命令行中设置，并从外部引入一个 Java 程序。这就是本节的内容。网络服务器运行在不同的端口上，在开发模式下通常运行在端口 8080 上。

任务：

- 编写一个可以接收来自不同来源端口信息的程序：
  在命令行中，可以用 --port=8000 指定端口。
  如果 --port 不存在，则应评估环境变量 port，该变量也可以通过命令行参数 dport=8020 设置。
  如果环境变量不存在，则应在 application.properties 文件中评估诸如 port=8888 之类的赋值。
  如果完全没有指定，则端口号默认设置为 8080。
- Zum Schluss soll der Port ausgegeben werden.
  最后，端口会被输出。

例如 3 个呼叫的变量：

- $ java com.tutego.exercise.os.PortConfiguration
- $ java com.tutego.exercise.os.PortConfiguration --port=8000
- $ java -Dport=8020 com.tutego.exercise.os.PortConfiguration

10.3 运行外部程序

作为一种独立于平台的编程语言，Java 不可能提供一切，因此它有各种方式向主机环境、操作系统提出请求。一个简单的方法是调用外部程序。这就是下一个任务所要讲的。

10.3.1 通过 Windows Management Instrumentation 读取电池状态

Bonny Brain 正用她的笔记本计算机玩一个战略游戏。下一轮比赛将持续 30 分钟，由于她的笔记本计算机只靠电池供电，所以她不希望在电池电量即将耗尽时停止游戏。通过以下调用，可以在 Windows 操作系统中通过命令行确定剩余运行时间的百分比和根据当前负载估计的剩余分钟数：

wmic path win32_battery get EstimatedChargeRemaining
```
EstimatedChargeRemaining
10
```

```
wmic path win32_battery get EstimatedRunTime
```
```
EstimatedRunTime
37
```

如果计算机不是带电池的笔记本计算机，则输出为"无可用实例"。当笔记本计算机充电时，EstimatedRunTime 的结果为 71 582 788（十六进制 4444444）。微软公司在 https://docs.microsoft.com/ en-us/windows/win32/cimwin32prov/win32-battery 上提供了这方面的详细信息。

任务：
- 通过 ProcessBuilder 启动 Windows 程序 wmic 作为 Java 程序的外部进程，并读出 EstimatedChargeRemaining 和 EstimatedRunTime 的结果。
- 在程序和输出中，要考虑到笔记本计算机可以通过电源运行，或者台式计算机没有电池。

信息：
Windows 操作系统提供了一个名为 Windows Management Instrumentation（WMI）的 API，可用于读取和部分修改计算机的设置。WMI 在 Java 方面类似 JMX。由于 Shell 脚本通常希望访问此任务以实现自动化，所以微软公司创建了 Windows Management Instrumentation（WMIC）命令行界面。

WMI 提供有关 CPU 使用率、主板、网络、电池状态等的信息。以下调用提供了一个小的概述：

```
wmic /?
```

10.4 建议解决方案

任务 10.1.1：彩色的控制台输出

```
private static final int EOF = -1;
private static final int HEX_PER_LINE = 32;
private static void printColorfulHexDump( Path path ) throws IOException {
  try ( InputStream is = new BufferedInputStream(
      Files.newInputStream( path ) ) ) {
    for ( int i = 0, b; (b = is.read()) != EOF; i++ ) {
      String color = b == 0 ? ANSI_GREEN :
```

```
                    b == 0xFF ? ANSI_RED :
                    Character.isDigit( b ) ? ANSI_PURPLE :
                    Character.isLetter( b ) ? ANSI_BLUE :
                    b == ' ' ? ANSI_CYAN :
                    ANSI_RESET;
        System.out.printf( "%s%02X ", color, b );
        if ( i % HEX_PER_LINE == (HEX_PER_LINE - 1) )
          System.out.println();
      }
    }
  }
}
```

源代码 10.1 com/tutego/exercise/os/AnsiColorHexDumper.java

我们不是用 Files.readAllBytes(...) 一次性读取输入，而是打开一个 InputStream，然后逐个字节读取。for 循环声明了两个变量。变量 i 控制换行符，它应该总是在 32 个十六进制字符之后被设置。变量 b 包含读取的字节。只要结果等于 -1，即还有可用的字节，就会一直读取 InputStream。

在 for 循环的主体中，被读取的字节会被测试，根据赋值情况，变量 color 是用 ANSI 转义序列初始化的。如果该字节不属于这 5 个类别，则颜色会被重置。最后，转义序列与十六进制代码一起被写入，在一行中最多有 32 个十六进制字符，然后就会设置换行。

任务 10.2.1: Windows 或者 UNIX 或者 macOS？

```
public enum OS {
  WINDOWS,
  MACOS,
  UNIX,
  UNKNOWN;

  public static OS current() {
    String osName = System.getProperty( "os.name" );
    if ( osName == null ) return UNKNOWN;
    osName = osName.toLowerCase();
    return osName.contains( "windows" ) ? OS.WINDOWS :
           osName.contains( "mac" ) ? OS.MACOS :
```

```
        osName.contains( "nix" ) || osName.contains( "nux" ) ? OS.UNIX :
        UNKNOWN;
  }
}
```

源代码 10.2 com/tutego/exercise/os/OS.java

current( ) 方法从特性 os.name 中读取操作系统的名称。如果结果为 null，则答案很快就会清楚，否则该方法将比较的名称转换为小写字母，并使用条件运算符级联测试各种子字符串。如果没有发现大小写区别，则操作系统的名称是未知的。

任务 10.2.2：统一命令行特性和文件中的特性

由于所有的参数，不管它们是如何呈现的，都是字符串，所以有一个单独的方法 parseInt(String)，它会将一个字符串转换成一个数字：

```
private static final String PORT = "port";
private static final int DEFAULT_PORT = 8080;

private static OptionalInt parseInt( String value ) {
  try {
    return OptionalInt.of( Integer.parseInt( value ) );
  }
  catch ( NumberFormatException e ) {
    return OptionalInt.empty();
  }
}
```

源代码 10.3 com/tutego/exercise/os/PortConfiguration.java

方法 parseInt(String) 访问 Integer.parseInt(...)，如果字符串不能被转换为数字，则提示异常。我们的方法捕捉到了这个异常，并返回一个 OptionalInt.empty( )，否则就是一个带有解析数字的 OptionalInt。

有下面的两个辅助方法：

```
private static OptionalInt portFromCommandLine( String[] args ) {
  for ( String arg : args )
    if ( arg.startsWith( "--" + PORT + "=" ) )
```

```
      return parseInt( arg.substring( ("--" + PORT + "=").length() ) );
    return OptionalInt.empty();
  }

  private static OptionalInt portFromPropertyFile() {
    String filename = "/application.properties";
    try ( InputStream is = PortConfiguration.class.getResourceAsStream(
          filename ) ) {
      Properties properties = new Properties();
      properties.load( is );
      return parseInt( properties.getProperty( PORT ) );
    }
    catch ( IOException e ) { /* Ignore */ }
    return OptionalInt.empty();
  }
```

源代码 10.4 com/tutego/exercise/os/PortConfiguration.java

portFromCommandLine(...) 接收命令行参数作为评估的参数。由于原则上可以向程序传递几个命令行参数，所以该方法会遍历所有控制台传递的参数，并查看是否包括 --port=。如果包含，则该方法会切断前面的部分并返回 parseInt(String) 的结果。如果该方法没有找到任何东西，则 OptionInt 为空 ()。

portFromPropertyFile(...) 解析类路径中的 application.properties 文件。类路径中的文件内容是通过类方法 getResourceAsStream(...) 读入的，因为它也支持打包在 JAR 存档中的资源。我们首先建立一个空的 Properties 对象，然后调用 load(...) 方法并传递 InputStream。如果文件不存在，则 InputStream 将为空，这会抛出异常，但 catch 块会捕获它，以及可能的加载过程中错误。如果 Properties 对象可以被填充，则我们询问端口特性并通过我们自己的方法将其转换为 OptionalInt；如果它被填充，则返回该值。

最后一种方法是 port(...)，它将所有内容联系在一起：

```
static int port( String[] args ) {
  // Step 1
  OptionalInt maybePort = portFromCommandLine( args );
  if ( maybePort.isPresent() )
    return maybePort.getAsInt();
```

```java
  // Step 2
  OptionalInt maybePortProperty = parseInt( System.getProperty( PORT ) );
  if ( maybePortProperty.isPresent() )
    return maybePortProperty.getAsInt();

  // Step 3
  OptionalInt maybePortApplicationProperty = portFromPropertyFile();
  if ( maybePortApplicationProperty.isPresent() )
    return maybePortApplicationProperty.getAsInt();

  // Step 4
  return DEFAULT_PORT;
}
```

源代码 10.5 com/tutego/exercise/os/PortConfiguration.java

首先测试命令行。如果匹配，则我们不必考虑其他任何事情，因为在命令行中传递具有最高优先级。

如果第一步没有退出，则继续第二步。System.getProperty(String) 读取系统特性，我们在 parseInt(...) 中传递返回值。如果该特性不存在，则 parseInt(...) 返回 empty( )。如果有赋值，则 port(...) 将其返回。

如果没有具有关联值的密钥，则我们需要打开该文件，否则将进入第三步。

如果没有文件或端口规范，则转到最后一步，即第四步。我们没有更多选项，并返回 8080 作为默认端口。

任务 10.3.1: 通过 Windows Management Instrumentation 读取电池状态

```java
static OptionalInt wmicBattery( String name ) {
  try {
    String[] command = { "CMD", "/C", "wmic", "path", "win32_battery",
                         "get", name };
    Process process = new ProcessBuilder( command ).start();
    try ( InputStream is = process.getInputStream();
          Reader isr = new InputStreamReader( is );
          Stream<String> stream = new BufferedReader( isr ).lines() ) {
      return stream.map( String::trim )
                   .filter( s -> s.matches( "\\d+" ) )
```

```
                    .mapToInt( Integer::parseInt )
                    .findFirst();
    }
  }
  catch ( IOException e ) {
    Logger.getLogger( WmicBattery.class.getName() ).info( e.toString() );
    return OptionalInt.empty();
  }
}
```
源代码 10.6 com/tutego/exercise/os/WmicBattery.java

由于我们的程序使用不同的参数调用 WMIC 两次，所以建议解决方案引入了一个新方法 wmicBattery(...)，在我们的例子中它可以传递 EstimatedChargeRemaining 和 EstimatedRunTime。这很有效，因为这两个查询总是会有一个方法 wmicBattery(...) 声明一个新数组并将其传递给 ProcessBuilder 的构造函数，以便可以使用 start( ) 启动该进程。方法返回的是我们在下一步中获取输出的 Process，也就是 InputStream。我们不必自己向进程中写入任何内容，因此不需要单独的 OutputStream，也不必等待程序结束，它会自动结束。

有不同的方法从 InputStream 中提取相关部分。建议解决方案将 InputStream 转换为一个 Buffered- Reader，这样就可以使用 lines( ) 方法，该方法返回一个所有行的流。这个数据流首先删除了每节的开头和结尾的空白部分。在这个流中，首先去除每行开头和结尾的空白，仅保留包含数字的行。结果要么剩下一行，要么没有任何行。如果一行包含一个数字，则它将被转换，并创建一个 IntStream，其中 findFirst( ) 带给我们一个 OptionalInt，它要么是空的（例如，由于缺少电池，进程输出中没有数字），要么包含数字。拦截并记录处理中可能的错误，然后返回一个 OptionalInt.emtpy( )。

在使用中，该方法可以像下面这样调用：

```
wmicBattery( "EstimatedChargeRemaining" ).ifPresentOrElse(
    value -> System.out.printf( "Estimated charge remaining:%d%%%n",
      value ),
    () -> System.out.println( "No instances available." ) );

wmicBattery( "EstimatedRunTime" ).ifPresentOrElse(
    minutes -> System.out.printf( minutes == 0X4444444 ?
      "Charging" :
```

```
            "Estimated run time:%d:%02d h (%d min)%n",
            minutes / 60, minutes% 60, minutes ),
    () -> System.out.println( "No instances available." ) );
```
源代码 10.7 com/tutego/exercise/os/WmicBattery.java

使用 EstimatedChargeRemaining，没有需要特殊解释的值。如果 OptionalInt 是空的，则应该执行与 OptionalInt 包含内容不同的操作。这就是 Java 9 中引入的 ifPresentOrElse(IntConsumer action, Runnable emptyAction) 的方法的用处。旧库的用户将不得不重写这个方法。

使用 EstimatedRunTime 时有一种特殊情况，即当前正在为电池充电，因此无法估计剩余时间。该代码测试了这个情况，然后返回一个不同的字符串。参数总是被传递给 printf(...) 方法，在充电情况下这当然会被忽略。

> **提示:**
> 属性的输出很快就会变得混乱，如下所示：
>
> **wmic diskdrive get**
> Availability BytesPerSector Capabilities CapabilityDescriptions Caption CompressionMethod ConfigManagerErrorCode ConfigManagerUserConfig CreationClassName DefaultBlockSize Description DeviceID ErrorCleared ErrorDescription ErrorMethodology FirmwareRevision Index InstallDate InterfaceType LastErrorCode Manufacturer MaxBlockSize MaxMediaSize MediaLoaded MediaType MinBlockSize Model Name NeedsCleaning NumberOfMediaSupported Partitions PNPDeviceID PowerManagementCapabilities PowerManagementSupported SCSIBus SCSILogicalUnit SCSIPort SCSITargetId
> SectorsPerTrack SerialNumber Signature Size Status StatusInfo SystemCreationClassName SystemName TotalCylinders TotalHeads TotalSectors TotalTracks TracksPerCylinder
> 512 {3, 4} {"Random Access", "Supports Writing"} SAMSUNG MZVLB1T0HALR-00000 0 FALSE Win32_DiskDrive Laufwerk \\.\PHYSICALDRIVE1 EXA7201Q 1 SCSI (Standardlaufwerke) TRUE Fixed hard disk media SAMSUNG MZVLB1T0HALR-00000
> \\.\PHYSICALDRIVE1 3 SCSI\DISK&VEN_NVME&PROD_SAMSUNG_MZVLB1T0\ 5&1E3C5E74&0&000000 0 0 1 0 63 0025_3886_81B2_A1AD. 1024203640320

OK Win32_
ComputerSystem DESKTOP-0P7C7G7 124519 255 2000397735 31752345 255 512 {3, 4, 10} {"Random Access", "Supports Writing", "SMART Notification"} WDC WD40EZRX-22SPEB0 0 FALSE Win32_DiskDrive Laufwerk \\.\PHYSICALDRIVE0 80.00A80 0 IDE (Standardlaufwerke) TRUE Fixed hard disk media WDC WD40EZRX- 22SPEB0 \\.\PHYSICALDRIVE0 1 SCSI\DISK&VEN_WDC&PROD_WD40EZRX-22SPEB0\4&2D010F8D&0&000000 0 0 0 0 63 WD-WCC4E2SHPE5N 4000784417280 OK Win32_
ComputerSystem DESKTOP-0P7C7G7 486401 255 7814032065 124032255 255

Dies ist schwer zu parsen, weshalb WMIC unterschiedliche Ausgabeformate bietet. Am Ende des Kommandos lässt sich ein /format:csv anhängen, das die Ausgabe deutlich einfacher zu verarbeiten macht.

wmic diskdrive get /format:csv

Node,Availability,BytesPerSector,Capabilities,CapabilityDescriptions,Captio n,CompressionMethod,ConfigManagerErrorCode,ConfigManagerUserConfig,Creation ClassName,DefaultBlockSize,Description,DeviceID,ErrorCleared,ErrorDescripti on,ErrorMethodology,FirmwareRevision,Index,InstallDate,InterfaceType,LastEr rorCode,Manufacturer,MaxBlockSize,MaxMediaSize,MediaLoaded,MediaType,MinBlo ckSize,Model,Name,NeedsCleaning,NumberOfMediaSupported,Partitions,PNPDevice ID,PowerManagementCapabilities,PowerManagementSupported,SCSIBus,SCSILogical Unit,SCSIPort,SCSITargetId,SectorsPerTrack,SerialNumber,Signature,Size,Stat us,StatusInfo,SystemCreationClassName,SystemName,TotalCylinders,TotalHeads, TotalSectors,TotalTracks,TracksPerCylinder
DESKTOP-0P7C7G7,,512,{3;4},{Random Access;Supports Writing},SAMSUNG MZVLB1T0HALR-00000,,0,FALSE,Win32_DiskDrive,,Laufwerk,\\.\ PHYSICALDRIVE1,,,,EXA7201Q,1,,SCSI,,(Standardlaufwerke),,,TRUE,Fixed hard disk media,,SAMSUNG MZVLB1T0HALR-00000,\\.\PHYSICALDRIVE1,,,3,SCSI\DISK
&VEN_NVME&PROD_SAMSUNG_MZVLB1T0\5&1E3C5E74&0 &000000,,,0,0,1,0,63,0025_3886_81B2_A1AD.,,1024203640320,OK,,Win32_ ComputerSystem,DESKTOP-0P7C7G7,124519,255,2000397735,31752345,255

```
DESKTOP-0P7C7G7,,512,{3;4;10},{Random Access;Supports
Writing;SMART Notification},WDC WD40EZRX-22SPEB0,,0,FALSE,Win32_
DiskDrive,,Laufwerk,\\.\ PHYSICALDRIVE0,,,,80.00A80,0,,IDE,,(S
tandardlaufwerke),,,TRUE,Fixed hard disk media,,WDC WD40EZRX-
22SPEB0,\\.\PHYSICALDRIVE0,,,1,SCSI\DISK&amp;VEN_ WDC&amp;PROD_
WD40EZRX-22SPEB0\4&amp;2D010F8D&amp;0&amp;000000,,,0,0,0,0,63,
WD-WCC4E2SHPE5N,,4000784417280,OK,,Win32_ComputerSystem,DESKTOP-
0P7C7G7,486401,255,7814032065,124032255,255
```

In der ersten Zeile stehen die Spaltennamen, und es folgen kommasepariert die Werte.

Falls nur ausgewählte Schlüssel gefragt sind, werden sie hinter get aufgeführt, etwa:

wmic diskdrive get Model,Size,SerialNumber /format:csv

```
Node,Model,SerialNumber,Size
DESKTOP-0P7C7G7,SAMSUNG MZVLB1T0HALR-00000,0025_3886_81B2_
A1AD.,1024203640320 DESKTOP-0P7C7G7,WDC WD40EZRX-22SPEB0, WD-
WCC4E2SHPE5N,4000784417280
```

Auch /format:list ist gut zu parsen, denn es entsteht das unter Java bekannte Prop- erty-Format, was dann praktisch ist, wenn es ein Ergebnis gibt. Es gibt viele weitere Formate.

第 11 章
反射、注解和 JavaBeans

Reflection 为我们提供了观察一个正在运行的 Java 程序的可能性。我们可以查询一个类有什么属性，然后在任何对象上调用方法，并读取和修改对象或类的变量。许多框架都使用了反射 API，例如用于对象关系映射的 JPA 或用于将 Java 对象映射到 XML 树的 JAXB。我们将自己编写一些带有 Reflection 的示例。

注释是一种自编程修改器。我们可以通过注解元数据来详细说明源代码，这些元数据可以在以后通过 Reflection 或其他工具读取。通常我们只是阅读别人的注释，但在本章中，我们还将练习如何编写自己的注释。

本章使用的数据类型如下：

- java.lang.Class (https://docs.oracle.com/en/java/javase/11/docs/api/java.base/java/lang/Class.html)
- java.lang.reflect.Field (https://docs.oracle.com/en/java/javase/11/docs/api/ java.base/java/lang/reflect/Field.html)
- java.lang.reflect.Method (https://docs.oracle.com/en/java/javase/11/docs/api/ java.base/java/lang/reflect/Method.html)
- java.lang.reflect.Constructor (https://docs.oracle.com/en/java/javase/11/docs/api/java.base/java/lang/reflect/Constructor.html)
- java.lang.reflect.Modifier (https://docs.oracle.com/en/java/javase/11/docs/api/java.base/java/lang/reflect/Modifier.html)

11.1 反射 API

反射 API 可以用来检查任何对象，下面的任务使用它创建任何数据类型的 UML 图示。这些任务侧重于实际的应用领域。你也可以用反射 API 做很多无意义的事情，例如改变不可变字符串的字符，但这很无聊，我们不想这样做。

11.1.1　创建具有继承关系的 UML 图示 ★

UML 图示在系统的文档中非常实用。一些 UML 图示也可以由工具自动生成。我们想自己编写一个这样的工具。起点是任何类，该类由反射 API 检查。我们可以读出这个类的所有属性，并生成一个 UML 图示。

由于 UML 图示是图形化的，所以问题自然产生了：如何在 Java 中绘制图形？我们不想解决这个问题，而是使用 PlantUML 描述语言（https://plantuml.com/）。PlantUML 之于 UML 图示来说，就像 HTML 之于网页和 SVG 之于矢量图。

示例：
```
interface Serializable << interface >>
Radio ..|> Serializable
ElectronicDevice --|> Radio
```

这里，箭头 --|> 或 <|-- 有规律地显示，..|> 或 <|.. 表示虚线。

从这些文本文档中，PlantUML 生成以下类型的表示（见图 11.1）。

PlantUML 是开源的，可以安装一个命令行实用程序，将文本描述转换为带有 UML 图示的图形。还有像 https://www.planttext.com/ 这样的网站可以实时呈现 UML 图示，见图 11.1。

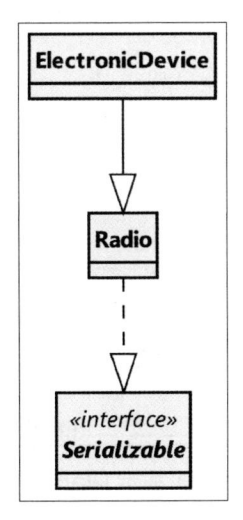

图 11.1　用图形表示 PlantUML 的语法

任务：

对于仅给出完全限定名称的任何类，生成 PlantUML 图表文本并将文本输出到控制台。

- 该图形应显示该类型及其基类型（超类和实现的接口）
- 该图形还应该递归地列出超类的类型

示例：

对于 Class.forName("java.awt.Point")，输出可能看起来像这样：

```
Point2D <|-- Point
Object <|-- Point2D
Cloneable <|.. Point2D
Serializable <|.. Point
```

11.1.2　创建带有属性的 UML 图示 ★

PlantUML 不仅可以描述类型关系（例如继承、接口和关联的实现），还可以描述对象/类变量和方法：

```
class Radio {
isOn: boolean
isOn() : boolean
{static} format(number: int): String
}
```

结果见图 11.2。

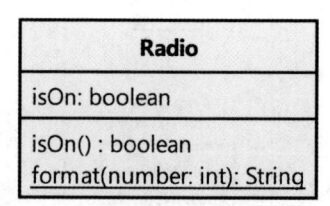

图 11.2　用图形表示 PlantUML 的语法

任务：

- 用 PlantUML 编写一个获取任何类对象并返回多行字符串作为结果的方法。

▶ 只包括对象 / 类的变量、构造函数和方法，不包括类型关系。

示例：
对于 java.awt.Dimension 类，输出可能看起来像这样：

```
@startuml
class Dimension {
    + width: int
    + height: int
    -  serialVersionUID: long
    + Dimension(arg0: Dimension)
    + Dimension()
    + Dimension(arg0: int, arg1: int)
    + equals(arg0: Object): boolean
    + toString(): String
    + hashCode(): int
    + getSize(): Dimension
    - initIDs(): void
    + setSize(arg0: Dimension): void
    + setSize(arg0: double, arg1: double): void
    + setSize(arg0: int, arg1: int): void
    + getWidth(): double
    + getHeight(): double
}
@enduml
```

11.1.3 从清单条目中生成 CSV 文件

在 CSV 文件中，条目由逗号或分号分隔，看起来像这样：

```
1;2;3
4;5;6
```

任务：

▶ 编写一个静态方法 writeAsCsv(List<?> objects, Appendable out)，该方法运行列表中的所有对象，通过反射 API 提取所有信息，然后将结果以 CSV 格式写入传递的 OutputStream。

- 为了提取信息，我们可以调用公共的 Bean-Getter（如果我们想了解属性）或访问（内部）对象变量。解决方案可以使用这两个变量中的任何一个。

示例：
```
Point p = new Point( 1, 2 );
Point q = new Point( 3, 4 );
List<?> list = Arrays.asList( p, q );
Writer out = new StringWriter();
writeAsCsv( list, out );
System.out.println( out );
```

奖励：
如果使用对对象变量的访问，则标有修饰符 transient 的对象变量不应该被写入。

11.2 注释

通过注解，我们可以在 Java 代码中引入元数据，以后可以读出来（通常是通过反射 API）。注释已经变得非常重要，因为现在的开发者以声明的方式表达了很多东西，而把实际的执行留给了框架。

11.2.1 从注释的对象变量中创建 CSV 文件★★

下面给出的是一个带有注解的类：

```
@Csv
class Pirate {
  @CsvColumn String name;
  @CsvColumn String profession;
  @CsvColumn int height;
  @CsvColumn( format = "### €" ) double income;
  @CsvColumn( format = "###.00" ) Object weight;
  String secrets;
}
```

任务：
- 声明注解 @Csv，它只能在类型声明中被设置。

- 声明注解 @CsvColumn，它只能被设置在对象变量上。
- 对于 @CsvColumn，允许一个字符串属性格式，通过 DecimalFormat 模式控制数字的格式。
- 创建一个 CsvWriter 类，其构造函数将类对象记为类型标记，以及稍后将写入 CSV 行的写入器。CsvWriter 类可以是 AutoCloseable。
- 将 CsvWriter 创建为泛型类型 CsvWriter<T>。
- Schreibe zwei neue Methoden 编写 2 个新的方法：
  - 空白的 writeObject(T object)：写入一个对象；
  - 空白的 write(Iterable<? extends T> iterable)：写入多个对象。
- CSV 列的分隔符默认为 ";"，但可以通过 delimiter(char) 方法改变。
- 考虑哪些错误情况可能发生并将其作为未检查的异常报告。

示例：
```
Pirate p1 = new Pirate();
p1.name = "Hotzenplotz";
p1.profession = null;
p1.height = 192;
p1.income = 124234.3234;
p1.weight = 89.10;
p1.secrets = "kinky";

StringWriter writer = new StringWriter();
try ( CsvWriter<Pirate> csvWriter =
         new CsvWriter<>( Pirate.class, writer ).delimiter( ',' ) ) {
  csvWriter.writeObject( p1 );
  csvWriter.writeObject( p1 );
}
System.out.println( writer );
```

11.3 建议解决方案

任务 11.1.1：创建具有继承关系的 UML 图示

```
public static void visitType( Class<?> clazz ) {

  if ( clazz.getSuperclass() != null ) {
```

```java
      System.out.printf( "%s <|-- %s%n",
                         clazz.getSuperclass().getSimpleName(),
                         clazz.getSimpleName() );
      visitType( clazz.getSuperclass() );
    }

    for ( Class<?> interfaze : clazz.getInterfaces() ) {
      System.out.printf( "%s <|.. %s%n",
                         interfaze.getSimpleName(),
                         clazz.getSimpleName() );
      visitType( interfaze );
    }
  }

  public static void main( String[] args ) throws ClassNotFoundException {
    Class<?> clazz = Class.forName( "javax.swing.JButton" );
    visitType( clazz );
    System.out.println( "hide members" );
  }
```

源代码 11.1 com/tutego/exercise/lang/reflect/PlantUmlTypeHierarchy.java

解决方案的核心是本身的方法 visitType(Class<?> clazz)，它被递归地调用。我们不必担心对象/类的变量、方法和构造函数，而只需要担心可能的超类和实现的接口。

第一部分区分处理可能的超类（最多可以有其中之一）。我们在类对象上调用 getSuperclass( )，如果已经到达继承层次结构中的 java.lang.Object 或超类，则获取 null。如果我们有一个超类，则程序会生成一个箭头。

当我们找到一个超类后，它也会再次有一个超类，因此我们递归地调用 visitType(...) 方法。

第二部分为实现的接口生成箭头。扩展的 for 循环遍历所有接口。如果该类没有实现一个接口，循环就不会被执行。如果有一个接口，那么我们会询问其名称以及我们自己的类的名称，并生成箭头。由于这个接口可以扩展其他接口，所以我们这里也递归地调用 visitType(...)。

任务 11.1.2: 创建带有属性的 UML 图示

```java
private static String plantUml( Class<?> clazz ) {

  StringWriter result = new StringWriter( 1024 );
  PrintWriter body = new PrintWriter( result );
  body.printf( "@startuml%nclass%s {%n", clazz.getSimpleName() );

  for ( Field field : clazz.getDeclaredFields() ) {
    String visibility = formatUmlVisibility( field );
    String type = field.getType().getSimpleName();
    body.printf( "%s%s:%s%n", visibility, field.getName(), type );
  }

  for ( Constructor<?> method : clazz.getConstructors() ) {
    String visibility = formatUmlVisibility( method );
    String parameters = formatParameters( method.getParameters() );
    body.printf( "%s %s(%s)%n",
      visibility, clazz.getSimpleName(), parameters );
  }

  for ( Method method : clazz.getDeclaredMethods() ) {
    String visibility = formatUmlVisibility( method );
    String parameters = formatParameters( method.getParameters() );
    String returnType = method.getReturnType().getSimpleName();
    body.printf( " %s%s(%s):%s%n",
      visibility, method.getName(), parameters, returnType );
  }

  body.println( "}\n@enduml" );
  return result.toString();
}

private static String formatParameters( Parameter[] parameters ) {
  return Arrays.stream( parameters )
           .map( p -> p.getName() + ": " + p.getType().getSimpleName() )
```

```
        .collect( Collectors.joining( ", " ) );
}

private static String formatUmlVisibility( Member field ) {
  return Modifier.isPrivate( field.getModifiers() ) ? "-" :
      Modifier.isPublic( field.getModifiers() ) ? "+" :
      Modifier.isProtected( field.getModifiers() ) ? "#" :
      "~";
}
```

源代码 11.2 com/tutego/exercise/lang/PlantUmlClassMembers.java

重点是方法 plantUml(Class)，它为一个 Class 对象生成 UML 图示的文本。当我们需要建立一个字符串时，有多种可能性。可以使用 +、- 运算符来连接字符串，或者使用 StringBuilder 和 append(...) 方法，如果使用格式化的字符串，那就更方便了。我们已经知道在 System.out 中类似 API 形式的 PrintWriter 类可以在这里使用。PrintWriter 为我们提供了不错的 print(...)，println(...) 和 printf(...) 方法。PrintWriter 是一个适配器，它需要一个目标来将生成的字符串写入其中。我们可以在一个 StringWriter 中收集结果。

然后，我们开始写 UML 图示的各部分。首先是类名，放入对象 / 类的变量，然后是构造函数，最后是方法。类对象通过相应的方法向我们提供数据。对象 / 类的变量、构造函数和方法有一个共同点，即它们具有可见性。这可以通过基本类型 Member 进行查询。独立的方法 formatUmlVisibility(Member) 将修改器翻译成一个字符串，我们将其包含在可见性的 PlantUML 标识中。

构造函数和方法还有一个共同点：参数列表。因此，还有一个方法 formatParameters(Parameter[]) 可以从参数对象的数组中为 PlantUML 生成一个字符串。它是一个数组，因为一个方法或构造函数可以有多个参数，所以我们需要询问每个参数的名称和返回类型。这可以看作一个逐步的转变，而这正是流 API 的用武之地。首先，我们从阵列中生成一个流。然后，我们将参数对象转移到一个字符串。做法如下：该字符串由参数的名称、冒号和数据类型组成。在 map(...) 操作之后，一个 stream<string> 被创建。这些字符串被放在一起，以逗号分隔，形成一个大的结果，然后返回。

任务 11.1.3 从清单条目中生成 CSV 文件

```
public static void writeAsCsv(
    List<?> objects, Appendable out ) throws IOException {
```

```java
    for ( Object object : objects ) {
      String line =
          Arrays.stream( object.getClass().getFields() )
               .filter( f -> ! Modifier.isTransient( f.getModifiers() ) )
               .map( f -> accessField( f, object ) )
               .collect( Collectors.joining( ";" ) );
      out.append( line ).append( "\n" );
    }
  }

  private static String accessField( Field field, Object object ) {
    try {
      return field.get( object ).toString();
    }
    catch ( IllegalAccessException e ) {
      throw new RuntimeException( e );
    }
  }
}
```

源代码 11.3 com/tutego/exercise/lang/reflect/ReflectionCsvExporter.java

该解决方案由所需的方法 writeAsCsv(List<?> objects, Appendable out) 和一个辅助方法 accessField(...) 组成。因为我们有一个具有任意元素的列表，所以首先要浏览这个列表。这里适合使用扩展的 for 循环。

当我们从列表中查看每个对象时，需要为每个对象生成一行。从一个对象到 CSV 行的这一连串操作是由流 API 实现的。如果以一个对象为起点，那么我们首先请求 Class 对象，然后用 getFields( ) 请求所有对象/类的变量。其结果是一个数组，我们要将其升级为一个流。因此，流由字段对象组成，由于我们不想考虑瞬时字段，所以使用流的过滤方法，只留下流中非瞬时的对象/类变量。

在下一步，我们必须读出对象变量。为此，我们采用了一种单独的方法。原因是 Lambda 表达式和已检查异常在语法上是无序的，因为检查性异常必须在 Lambda 表达式中捕获。然而，反射 API 经常使用检查过的异常。自定义方法 String accessField(Field,Object) 的目的是为一个给定的对象读取一个对象变量并将其转换为字符串。该方法捕获了一个可能的已检查的异常，并将其封装在一个未检查的异常中。流的终端操作收集所有部分字符串，并用分号将它们连接起来。最终，这一行（包括行尾符号）被写入 Writer。

任务 11.2.1: 从注释的对象变量中创建 CSV 文件

必须为解决方案声明两种注解类型。第一个类型 Csv 是 @Target(ElementType.TYPE)，因为它只能放在类型声明处，当然这个注解必须通过运行时（RUNTIME）读出来，RetentionPolicy.RUNTIME.Csv 是 @Target。

```
import java.lang.annotation.ElementType;
import java.lang.annotation.Retention;
import java.lang.annotation.RetentionPolicy;
import java.lang.annotation.Target;

@Retention( RetentionPolicy.RUNTIME )
@Target( ElementType.TYPE )
public @interface Csv {
}
```

源代码 11.4 com/tutego/exercise/annotation/Csv.java

第二种注解类型 CsvColumn 仅用于类 / 对象的成员变量，因此将其 @Target 设置为 ElementType.FIELD。然而，这并不能将其限制为仅应用于对象成员变量。

该注解同样会在运行时被解析。CsvColumn 具有一个名为 format 的属性，用于存储格式化字符串，默认为空字符串。稍后我们不会对空字符串进行处理。

```
import java.lang.annotation.ElementType;
import java.lang.annotation.Retention;
import java.lang.annotation.RetentionPolicy;
import java.lang.annotation.Target;

@Retention( RetentionPolicy.RUNTIME )
@Target( ElementType.FIELD )
public @interface CsvColumn {
  String format() default "";
}
```

源 代 码 11.5 com/tutego/exercise/lang/reflect/PlantUmlTypeHierarchy.java

我们继续讨论广泛的 CsvWriter：

```java
public class CsvWriter<T> implements AutoCloseable {

  private final List<Field> fields;
  private final Class<?> clazz;
  private final Writer writer;
  private char delimiter = ';';

  public CsvWriter( Class<T> clazz, Writer writer ) {
    if ( ! clazz.isAnnotationPresent( Csv.class ) )
      throw new IllegalArgumentException(
          "Given class is not annotated with @Csv" );

    fields = Arrays.stream( clazz.getDeclaredFields() )
                .filter( field -> field.isAnnotationPresent(
                         CsvColumn.class ) )
                .collect( Collectors.toList() );

    if ( fields.isEmpty() )
      throw new IllegalArgumentException(
          "Class does not contain any @CsvColumn" );

    this.clazz = clazz;
    this.writer = Objects.requireNonNull( writer );
  }

  public CsvWriter<T> delimiter( char delimiter ) {
    this.delimiter = delimiter;
    return this;
  }

  public void write( Iterable<? extends T> iterable ) {
    iterable.forEach( this::writeObject );
  }

  public void writeObject( T object ) {
```

第 11 章　反射、注解和 JavaBeans ｜ 337

```java
      if ( ! clazz.isInstance( object ) )
        throw new IllegalArgumentException(
            "Argument is of type " + object.getClass().getSimpleName()
          + " but must be of type " + clazz.getSimpleName() );
      String line = fields.stream()
                          .map( field -> getFieldValue( object, field ) )
                          .collect( Collectors.joining( Character.toString(
                              delimiter ), "", "\n" ) );
      try {
        writer.write( line );
      }
      catch ( IOException e ) {
        throw new UncheckedIOException( e );
      }
    }

    private String getFieldValue( Object object, Field field ) {
      try {
        Object fieldValue = field.get( object );

        if ( fieldValue == null )
          return "";

        String format = field.getAnnotation( CsvColumn.class ).format();
        if ( format.trim().isEmpty() )
          return Objects.toString( fieldValue );

        if ( isNumericType( fieldValue ) )
          return new DecimalFormat( format ).format( fieldValue );

        throw new IllegalStateException( "Only numeric types can be formatted, but type was " + fieldValue.getClass().getSimpleName() );
      }
      catch ( IllegalAccessException e ) {
        throw new RuntimeException( e );
      }
```

```
  }

  private static boolean isNumericType( Object value ) {
    return Stream.of( Integer.class, Long.class, Double.class,
                      BigInteger.class, BigDecimal.class )
            .anyMatch( clazz -> clazz.isInstance( value ) );
  }

  @Override public void close() {
    try {
      writer.close();
    }
    catch ( IOException e ) {
      throw new UncheckedIOException( e );
    }
  }
}
```

源代码 11.6 com/tutego/exercise/annotation/CsvWriter.java

该构造函数接收一个 Class 对象和一个 Writer，并进行检查和预处理。

1. 构造函数首先执行一个测试，看以后要写的程序是否带有标记注释 Csv；如果不带，就会出现一个运行时异常。这样以后编程时就不必再重复这种检查。由于以后编程时类型总是相同的，而且出于性能的考虑，应该避免在运行时进行反射，所以构造函数会获取相应的字段对象并保存起来。只有用 @CsvColumn 注解的字段对象才被放在内部列表中。如果这个列表是空的，那就有一个例外，因为这将意味着没有什么可写的。构造函数还记住了类的对象，以便在以后的测试中使用，因为泛型只是编译器的一个技巧，并且在运行时可能替换不正确的类型。
2. 写入器在内部被记录，并且像往常一样，早期测试确保该写入器不为空。我们的类实现了 AutoCloseable，实现的 close( ) 方法关闭了底层写程序。这使 CsvWriter 可以在 try-with-resources 块中使用。

分隔符是列的 CSV 分隔符，可以重新分配 delimiter（char）。该方法返回当前的 CsvWriter，这样就可以很好地级联调用，形成一个流畅的 API。该方法并不测试分隔符是否合理。原则上，分界符可以是 "a" "\n" 或 "\u000"。

writeObject(T) 写一个单一的对象，write(Iterable<? extends T>) 方法可以很好地访问该对象。Iterable 提供的 forEach(...) 方法在所有数据上运行，并在每个单独的元素上调用 writeObject(T)。

首先，writeObject(T) 执行了一个类型检查，看参数是否与通过类对象在构造函数中声明的类型兼容。clazz.isInstance(...) 是一个动态的 instanceof 操作。如果类型不匹配，就会出现一个异常。在下一步中，该方法遍历所有字段元素，使用 getFieldValue(...) 获得对象变量赋值并转换为每个字段的字符串，最后将所有字符串与 CSV 输出的分隔符结合。在该行的末尾添加一个换行符。结果字符串被写入 Writer，可能的 IOException 被捕获并被转换为未检查的异常并抛出。getFieldValue(...) 是一个单独的方法，它隐藏了访问和格式化数字值的复杂性。

getFieldValue(...) 方法获得了数据的对象和对象变量的字段对象。字段方法 get(...) 返回存储在对象变量中的值。如果它是空的，就没有东西可写，我们返回空字符串。现在有两种可能性。在 CsvColumn 中，有一个带有格式化模式的属性格式和一个不带格式化模式的属性格式。

- 如果没有格式化模式或格式只由空白组成，那么将返回读取值的字符串表示。
- 如果有一个格式化模式，则 isNumericType(...) 检查属性中的值是否是数字。这个属性可以在构造函数中测试，但这样做有一个缺点：例如，类型是 Object，而在运行时后面有一个 double，那是完全可以的。这只能在运行时进行测试。对于数字属性，DecimalFormat 的 format(..) 方法返回字符串表示。

如果给定一个格式化模式，但类型不是数字，就会出现 IllegalStateException。

DecimalFormat 可以自动格式化不同的类型。有效类型通过单独的方法 isNumericType(...) 进行测试。这些类型包括 Integer, Long, Double, BigInteger, BigDecimal。

编后语

祝贺那些花时间完成任务、研究建议解决方案并坚持到最后的读者。虽然路途遥远，但现在他们已经为一个有前途的 Java 事业奠定了基础。但当然，这本书之后并没有停止。那些成功的人一遍又一遍地重复以下三个步骤。

1. 阅读书籍，学习博客条目，观看学习视频。
2. 自己编程并对建议解决方案提出批评性质疑。
3. 研究其他人的源代码，识别和学习模式。

此外，还有一些网站和平台定期发布新任务。一小部分免费网站如下。

- Code Golf Stack Exchange。
  Code Golf Stack Exchange 是一个面向编程竞赛的网站，旨在将编程作为一种娱乐活动，这是它官方介绍中的说法。有些任务也会用 Java 来完成，而许多任务则使用一些非常奇特的编程语言来完成，这些编程语言只是因为其紧凑的编写风格而被创造出来的 (https://codegolf.stackexchange.com/questions)。
- Project Euler。
  "Project Euler 是一系列具有挑战性的数学或编程问题，要解决这些问题不仅需要数学见解。"许多问题都来自数学领域，但是开发者们必须借助精心设计的算法来解决它们 (https://projecteuler.net/)。
- Daily Programmer。
  Reddit 是一个"互联网的首页"，由一系列称为"Subreddits"的论坛组成。其中一个 Subreddit 是 Daily Programmer，旨在通过每周的编程任务挑战具有各种技能水平的程序员。这些任务的难度各不相同 (https://www.reddit.

com/r/dailyprogrammer/）。
- Rosetta Code。
Rosetta Code 的魅力在于其多样性的编程语言。该网站上有超过 1 000 个编程问题和 800 多种编程语言的解决方案。作为 Java 开发人员，我们可以通过其他编程语言的解决方案学习，虽然不太可能通过像 COBOL 这样的古老编程语言学习，但是可以通过函数式编程语言学到很多东西（https://rosettacode.org/wiki/Rosetta_Code）。

有些公司在招聘测试中使用编程任务。这种做法催生了一个商业模式，即商业提供者定期提供需要应聘者完成的封闭性任务。人力资源部门可以在后续的评估中与主要开发人员一起评估解决方案，并获取一些度量指标。

一个类似的发展是竞技编程。你像比赛一样开发解决方案。完成一个任务可以获得一定的分数，谁得分最高谁就是赢家。

附录 A
Java 领域中常见的类型和方法

一本关于 Java 编程任务的图书中，重要的是项目中相关的数据类型和方法。很少使用数据类型和方法几乎没有用处，特别是当重要的数据类型缺失时。在为这本书做准备时，我编写了一个软件，用于研究百余个开源库中出现的数据类型和方法，结果非常有趣。本附录展示了这些统计数据，其中所列出的是在实践中经常出现的类型。Java EE、Spring 或其他开源库的类型不在这里出现，这里只有 Java SE 的类型。

A.1 经常出现的类型的包

- java.io
  BufferedReader, ByteArrayInputStream, ByteArrayOutputStream, DataOutputStream, FileInputStream, FileOutputStream, File, IOException, InputStream, ObjectInputStream, ObjectOutputStream, OutputStream, PrintStream, PrintWriter, StringReader, StringWriter, Writer

- java.lang
  Appendable, AssertionError, Boolean, Byte, CharSequence, Character, ClassLoader, Class, Double, Enum, Exception, Float, IllegalArgumentException, IllegalStateException, IndexOutOfBoundsException, Integer, Iterable, Long, Math, NullPointerException, Number, Object, RuntimeException, Short, StringBuffer, StringBuilder, String, System, ThreadLocal, Thread, Throwable, UnsupportedOperationException

- java.lang.annotation
  Annotation

- java.lang.ref

SoftReference
▶ java.lang.reflect
　　Array, Constructor, Field, InvocationTargetException, Method, ParameterizedType
▶ java.math
　　BigInteger
▶ java.net
　　URI, URL
▶ java.nio
　　ByteBuffer
▶ java.security
　　AccessController
▶ java.sql
　　PreparedStatement
▶ java.text
　　MessageFormat
▶ java.util
　　AbstractList, ArrayList, Arrays, BitSet, Calendar, Collection, Collections, Comparator, Date, Enumeration, HashMap, HashSet, Hashtable, Iterator, LinkedHashMap, LinkedHashSet, LinkedList, List, Locale, Map, NoSuchElementException, Objects, Optional, Properties, ResourceBundle, Set, Stack, StringTokenizer, TreeMap, Vector
▶ java.util.concurrent
　　ConcurrentHashMap, ConcurrentMap
▶ java.util.concurrent.atomic
　　AtomicBoolean, AtomicInteger, AtomicLong, AtomicReference
▶ java.util.concurrent.locks
　　Lock, ReentrantLock
▶ java.util.function
　　Consumer, Function
▶ java.util.logging
　　Logger
▶ java.util.regex
　　Matcher, Pattern
▶ java.util.stream
　　Collectors, Stream

- javax.xml.bind
  JAXBElement
- javax.xml.namespace
  QName

A.2　100 个最常使用的类

表 A.1　经常使用的数据类型

类的名称	出现次数	百分比分布 /%
java.lang.StringBuilder	605.692	25.28
java.lang.String	190.854	7.97
java.lang.Object	138.093	5.76
java.util.Iterator	102.072	4.26
java.util.List	99.833	4.17
java.lang.StringBuffer	94.562	3.95
java.util.Map	69.415	2.90
java.lang.Class	52.435	2.19
java.util.ArrayList	46.855	1.96
java.lang.Integer	45.650	1.91
java.util.Set	35.131	1.47
java.util.HashMap	28.958	1.21
java.util.logging.Logger	26.173	1.09
java.lang.IllegalArgumentException	25.762	1.08
javax.xml.namespace.QName	22.050	0.92
java.io.File	19.951	0.83
java.lang.Boolean	19.919	0.83
java.lang.System	18.881	0.79
java.util.Collection	15.678	0.65

续表

类的名称	出现次数	百分比分布 /%
java.util.Arrays	15.533	0.65
java.lang.AssertionError	14.927	0.62
java.lang.IllegalStateException	14.189	0.59
java.util.Map$Entry	13.554	0.57
java.util.Collections	13.061	0.55
java.lang.Long	12.493	0.52
java.util.Hashtable	11.934	0.50
java.lang.Enum	11.191	0.47
java.lang.Math	10.670	0.45
java.lang.UnsupportedOperationException	10.434	0.44
java.lang.reflect.Method	10.357	0.43
java.io.PrintStream	10.278	0.43
java.util.Vector	10.118	0.42
java.lang.Character	9.904	0.41
java.lang.Thread	9.294	0.39
java.nio.ByteBuffer	9.285	0.39
java.util.HashSet	8.881	0.37
java.lang.Throwable	8.415	0.35
java.util.Properties	8.106	0.34
java.lang.Double	7.956	0.33
java.lang.IndexOutOfBoundsException	7.799	0.33
java.lang.RuntimeException	7.734	0.32
java.util.Objects	7.700	0.32
java.io.Writer	7.473	0.31

续表

类的名称	出现次数	百分比分布 /%
java.io.IOException	6.902	0.29
java.io.InputStream	6.539	0.27
java.util.stream.Stream	6.429	0.27
java.lang.CharSequence	6.170	0.26
java.lang.Exception	6.054	0.25
java.io.PrintWriter	5.785	0.24
java.math.BigInteger	5.520	0.23
java.util.Enumeration	5.047	0.21
java.util.Stack	4.904	0.20
java.util.ResourceBundle	4.777	0.20
java.io.OutputStream	4.682	0.20
java.util.LinkedList	4.646	0.19
java.util.Optional	4.169	0.17
java.io.ByteArrayOutputStream	4.115	0.17
java.util.AbstractList	4.074	0.17
java.lang.Float	4.062	0.17
java.util.StringTokenizer	4.024	0.17
java.lang.NullPointerException	3.840	0.16
java.util.concurrent.atomic.AtomicReference	3.579	0.15
java.net.URI	3.491	0.15
java.util.LinkedHashMap	3.490	0.15
java.lang.Iterable	3.483	0.15
java.lang.reflect.Field	3.452	0.14
java.lang.Number	3.449	0.14

续表

类的名称	出现次数	百分比分布 /%
java.net.URL	3.436	0.14
java.util.regex.Pattern	3.386	0.14
java.util.regex.Matcher	3.343	0.14
java.util.Calendar	3.281	0.14
java.util.concurrent.ConcurrentHashMap	3.231	0.13
java.text.MessageFormat	3.070	0.13
javax.xml.stream.XMLStreamReader	3.032	0.13
java.util.concurrent.locks.Lock	3.012	0.13
java.lang.ClassLoader	2.861	0.12
java.util.concurrent.atomic.AtomicInteger	2.835	0.12
java.lang.ThreadLocal	2.706	0.11
java.security.AccessController	2.687	0.11
java.util.concurrent.ConcurrentMap	2.579	0.11
java.util.BitSet	2.553	0.11
java.math.BigDecimal	2.437	0.10
java.sql.ResultSet	2.408	0.10
java.io.BufferedReader	2.359	0.10
java.io.DataOutputStream	2.264	0.09
java.util.concurrent.atomic.AtomicLong	2.245	0.09
java.sql.PreparedStatement	2.237	0.09
java.util.LinkedHashSet	2.222	0.09
java.util.Date	2.212	0.09
java.lang.ref.SoftReference	2.139	0.09

续表

类的名称	出现次数	百分比分布 /%
java.io.ObjectOutputStream	2.119	0.09
java.lang.reflect.Array	2.048	0.09
java.io.ObjectInputStream	2.043	0.09
java.io.StringWriter	1.992	0.08
java.util.concurrent.atomic.AtomicBoolean	1.991	0.08

A.3　100 种常用的方法

表 A.2　带有方法名称的类

方法名称	出现次数	百分比分布 /%
java.lang.StringBuilder#append	377.938	15.77
java.lang.StringBuilder#toString	111.511	4.65
java.lang.StringBuilder#<init>	111.272	4.64
java.lang.Object#<init>	80.559	3.36
java.lang.StringBuffer#append	61.081	2.55
java.lang.String#equals	56.525	2.36
java.util.Iterator#next	50.355	2.10
java.util.Iterator#hasNext	50.093	2.09
java.util.ArrayList#<init>	28.413	1.19
java.lang.Integer#valueOf	27.554	1.15
java.lang.String#length	26.959	1.13
java.util.Map#put	26.388	1.10
java.util.List#add	26.041	1.09
java.lang.IllegalArgumentException#<init>	25.190	1.05

续表

方法名称	出现次数	百分比分布 /%
java.lang.Object#getClass	24.894	1.04
java.util.List#size	21.020	0.88
java.util.Map#get	18.625	0.78
java.util.List#iterator	18.422	0.77
javax.xml.namespace.QName#<init>	17.541	0.73
java.lang.String#substring	16.687	0.70
java.lang.StringBuffer#toString	15.771	0.66
java.lang.Object#equals	15.554	0.65
java.lang.StringBuffer#<init>	15.488	0.65
java.lang.Class#getName	15.238	0.64
java.lang.AssertionError#<init>	14.877	0.62
java.lang.IllegalStateException#<init>	13.984	0.58
java.lang.String#charAt	12.815	0.53
java.util.Set#iterator	12.788	0.53
java.util.logging.Logger#log	12.576	0.52
java.util.List#get	12.204	0.51
java.util.HashMap#<init>	12.185	0.51
java.lang.Boolean#valueOf	10.513	0.44
java.util.HashMap#put	10.470	0.44
java.lang.UnsupportedOperation-Exception#<init>	10.340	0.43
java.util.Set#add	9.598	0.40
java.lang.System#arraycopy	9.116	0.38
java.lang.String#startsWith	8.501	0.35
java.lang.Object#toString	8.418	0.35

续表

方法名称	出现次数	百分比分布 /%
java.io.PrintStream#println	8.240	0.34
java.lang.String#indexOf	8.053	0.34
java.lang.IndexOutOfBoundsException#<init>	7.792	0.33
java.util.Collection#iterator	7.486	0.31
java.lang.RuntimeException#<init>	7.414	0.31
java.lang.Long#valueOf	7.268	0.30
java.lang.Boolean#booleanValue	7.132	0.30
java.lang.String#valueOf	7.027	0.29
java.lang.Integer#intValue	7.006	0.29
java.util.Hashtable#put	6.884	0.29
java.util.Map$Entry#getValue	6.781	0.28
java.util.Map$Entry#getKey	6.634	0.28
java.util.ArrayList#add	6.581	0.27
java.util.logging.Logger#isLoggable	6.530	0.27
java.util.HashSet#<init>	6.527	0.27
java.lang.Object#hashCode	6.479	0.27
java.io.Writer#write	6.423	0.27
java.lang.Class#getClassLoader	6.378	0.27
java.util.Arrays#asList	6.083	0.25
java.lang.String#equalsIgnoreCase	6.070	0.25
java.util.List#toArray	5.872	0.25
java.lang.String#format	5.769	0.24
java.io.File#<init>	5.631	0.24
java.lang.Enum#<init>	5.352	0.22

续表

方法名称	出现次数	百分比分布 /%
java.lang.Enum#valueOf	5.322	0.22
java.lang.String#<init>	5.298	0.22
java.lang.String#trim	5.115	0.21
java.io.IOException#<init>	5.079	0.21
java.util.Objects#requireNonNull	4.767	0.20
java.util.List#isEmpty	4.748	0.20
java.lang.Class#isAssignableFrom	4.468	0.19
java.util.Map#entrySet	4.254	0.18
java.lang.Throwable#addSuppressed	4.100	0.17
java.lang.String#hashCode	3.984	0.17
java.lang.Math#min	3.964	0.17
java.util.AbstractList#<init>	3.943	0.16
java.util.Map#containsKey	3.905	0.16
java.lang.Class#desiredAssertionStatus	3.884	0.16
java.lang.NullPointerException#<init>	3.808	0.16
java.util.ArrayList#size	3.758	0.16
java.lang.String#endsWith	3.718	0.16
java.lang.Character#valueOf	3.714	0.16
java.util.Set#contains	3.698	0.15
java.io.InputStream#close	3.684	0.15
java.util.ResourceBundle#getString	3.670	0.15
java.lang.Thread#currentThread	3.488	0.15
java.lang.Double#valueOf	3.465	0.14
java.lang.Iterable#iterator	3.363	0.14

续表

方法名称	出现次数	百分比分布 /%
java.lang.Integer#parseInt	3.350	0.14
java.util.HashMap#get	3.319	0.14
java.lang.System#getProperty	3.292	0.14
java.util.Map#values	3.240	0.14
java.util.Arrays#fill	3.202	0.13
java.util.ArrayList#get	3.189	0.13
java.lang.Integer#<init>	3.181	0.13
java.lang.Math#max	2.960	0.12
java.io.OutputStream#write	2.951	0.12
java.util.Map#remove	2.909	0.12
java.lang.Class#forName	2.903	0.12
java.lang.String#replace	2.887	0.12

A.4　100 个最常用的方法，包括参数列表

表 A.3　带有方法名称和参数列表的类

方法名称和参数列表	出现次数	百分比分布 /%
java.lang.StringBuilder#append(String)	296.283	12.37
java.lang.StringBuilder#toString( )	111.511	4.65
java.lang.StringBuilder#<init>( )	102.410	4.27
java.lang.Object#<init>( )	80.559	3.36
java.lang.String#equals(Object)	56.525	2.36
java.util.Iterator#next( )	50.355	2.10
java.util.Iterator#hasNext( )	50.093	2.09
java.lang.StringBuffer#append(String)	48.916	2.04

续表

方法名称和参数列表	出现次数	百分比分布 /%
java.lang.StringBuilder#append(Object)	33.491	1.40
java.lang.Integer#valueOf(int)	27.170	1.13
java.lang.String#length( )	26.959	1.13
java.util.Map#put(Object, Object)	26.388	1.10
java.util.List#add(Object)	25.409	1.06
java.lang.Object#getClass( )	24.894	1.04
java.lang.IllegalArgumentException#<init>(String)	22.429	0.94
java.lang.StringBuilder#append(int)	21.752	0.91
java.util.ArrayList#<init>( )	21.117	0.88
java.util.List#size( )	21.020	0.88
java.util.Map#get(Object)	18.625	0.78
java.util.List#iterator( )	18.422	0.77
javax.xml.namespace.QName#<init>(String, String)	17.076	0.71
java.lang.StringBuilder#append(char	16.402	0.68
java.lang.StringBuffer#toString( )	15.771	0.66
java.lang.Object#equals(Object)	15.554	0.65
java.lang.Class#getName( )	15.238	0.64
java.lang.StringBuffer#<init>( )	13.318	0.56
java.lang.String#charAt(int)	12.815	0.53
java.util.Set#iterator( )	12.788	0.53
java.util.List#get(int)	12.204	0.51
java.lang.IllegalStateException#<init>(String)	10.673	0.45
java.lang.AssertionError#<init>( )	10.622	0.44

续表

方法名称和参数列表	出现次数	百分比分布 /%
java.util.HashMap#put(Object, Object)	10.470	0.44
java.util.HashMap#<init>( )	10.273	0.43
java.util.Set#add(Object)	9.598	0.40
java.lang.String#substring(int, int)	9.538	0.40
java.lang.Boolean#valueOf(boolean)	9.449	0.39
java.lang.System#arraycopy(Object, int, Object, int, int)	9.116	0.38
java.lang.Object#toString( )	8.418	0.35
java.lang.String#startsWith(String)	8.344	0.35
java.io.PrintStream#println(String)	7.568	0.32
java.util.Collection#iterator( )	7.486	0.31
java.lang.String#substring(int)	7.149	0.30
java.lang.Boolean#booleanValue( )	7.132	0.30
java.lang.Long#valueOf(long)	7.066	0.29
java.lang.Integer#intValue( )	7.006	0.29
java.lang.IndexOutOfBoundsException#<init>( )	6.941	0.29
java.util.Hashtable#put(Object, Object)	6.884	0.29
java.util.Map$Entry#getValue( )	6.781	0.28
java.util.Map$Entry#getKey( )	6.634	0.28
java.util.logging.Logger#isLog- gable(java.util.logging.Level)	6.530	0.27
java.lang.Object#hashCode( )	6.479	0.27
java.lang.Class#getClassLoader( )	6.378	0.27
java.util.ArrayList#add(Object)	6.359	0.27
java.util.Arrays#asList(Object…)	6.083	0.25

续表

方法名称和参数列表	出现次数	百分比分布 /%
java.lang.String#equalsIgnore-Case(String)	6.070	0.25
java.lang.UnsupportedOperation-Exception#<init>( )	5.769	0.24
java.util.List#toArray(Object…)	5.691	0.24
java.lang.Enum#<init>(String, int)	5.352	0.22
java.lang.Enum#valueOf(Class, String)	5.322	0.22
java.lang.String#trim( )	5.115	0.21
java.lang.StringBuffer#append(char)	5.074	0.21
java.lang.StringBuilder#append(long)	5.035	0.21
java.util.HashSet#<init>( )	5.015	0.21
java.util.ArrayList#<init>(int)	4.941	0.21
java.util.logging.Logger#log(java.util.logging.Level, String, Throwable)	4.766	0.20
java.util.List#isEmpty( )	4.748	0.20
java.io.Writer#write(String)	4.690	0.20
java.util.logging.Logger# log(java.util.logging.Level, String)	4.577	0.19
java.lang.StringBuilder#<init>(String)	4.512	0.19
java.lang.UnsupportedOperation-Exception#<init>(String)	4.490	0.19
java.lang.Class#isAssignable-From(Class)	4.468	0.19
java.lang.String#format(String, Object…)	4.411	0.18
java.io.IOException#<init>(String)	4.321	0.18
java.lang.StringBuilder#<init>(int)	4.314	0.18
java.util.Map#entrySet( )	4.254	0.18
java.lang.AssertionError#<init>(Object)	4.129	0.17

续表

方法名称和参数列表	出现次数	百分比分布 /%
java.lang.Throwable#addSuppressed(Throwable)	4.100	0.17
java.lang.String#indexOf(int)	4.088	0.17
java.lang.String#hashCode( )	3.984	0.17
java.util.AbstractList#<init>( )	3.943	0.16
java.util.Map#containsKey(Object)	3.905	0.16
java.lang.Class#desiredAssertion-Status( )	3.884	0.16
java.lang.RuntimeException#<init>(String)	3.801	0.16
java.util.ArrayList#size( )	3.758	0.16
java.lang.String#endsWith(String)	3.718	0.16
java.lang.Character#valueOf(char)	3.714	0.16
java.util.Set#contains(Object)	3.698	0.15
java.io.InputStream#close( )	3.684	0.15
java.util.ResourceBundle#getString(String)	3.670	0.15
java.lang.Thread#currentThread( )	3.488	0.15
java.lang.Iterable#iterator( )	3.363	0.14
java.util.HashMap#get(Object)	3.319	0.14
java.lang.String#valueOf(Object)	3.288	0.14
java.lang.StringBuffer#append(int)	3.248	0.14
java.util.Map#values( )	3.240	0.14
java.lang.Double#valueOf(double)	3.229	0.13
java.util.ArrayList#get(int)	3.189	0.13
java.lang.Integer#<init>(int)	3.120	0.13
java.util.Objects#requireNon-Null(Object, String)	3.102	0.13

《漫画学 Java》三部曲

别告诉我你不懂 Java！ 薛定谔来教你！

全彩印刷 零基础 漫画版

扫码购书